Ben Stacy Jerrik (Ed.)

Bohunice Nuclear Power Plant

Ben Stacy Jerrik (Ed.)

Bohunice Nuclear Power Plant

Nuclear reactor, Jaslovské Bohunice, Pressurized water reactor

Part Press

Publisher:
Part Press is a trademark of
International Book Market Service Ltd., 17 Rue Meldrum, Beau Bassin, 1713-01 Mauritius
Email: info@bookmarketservice.com
Website: www.bookmarketservice.com

Published in 2013

Printed in: U.S.A., U.K., Germany. This book was not produced in Mauritius.

ISBN: 978-620-1-57369-7

Contents

Articles

References

Bohunice_Nuclear_Power_Plant

The **Bohunice Nuclear Power Plant** (NPP) (Slovak: *Atómové elektrárne Bohunice*, abbr. EBO) is a complex of nuclear reactors situated 2.5 km from the village of Jaslovské Bohunice in the Trnava District in western Slovakia.

Bohunice NPP comprises two plants: V1 and V2. Both plants contain two reactor units. The plant was connected to the national power network in stages in the period between 1978 and 1985. The four power reactors are pressurized water reactors of the Soviet VVER-440 design.

Annual electricity generation averages about 12,000 GWh. Upon development of a district heating supply network in the town of Trnava near Bohunice NPP, V2 switched to co-generation. Part of this system is a heat feeder line commissioned in 1987. In 1997 a heat feeder line to Leopoldov and Hlohovec was begun, branching off from the Trnava line.

Bohunice A1 (shutdown)

The A1 is another nuclear reactor situated on the Jaslovské Bohunice site. On February 22, 1977 the A1 reactor suffered a major accident during refueling, rated INES-4. This reactor is currently undergoing a decommissioning and cleanup process.

Bohunice V1 (shutdown)

As a condition of accession into the European Union (2004) Slovakia was forced to deactivate the two reactors at the V1 plant. A provision in the accession treaty allowed for reactivation in case of an emergency.

V1 plant was exempted from Slovenské Elektrárne sale to Enel and transferred to JAVYS, a national nuclear decommissioning company fully owned by the state. The first reactor was shut down at the end of 2006, the second on the last day of 2008.

The Russia-Ukraine gas dispute in January 2009 disrupted natural gas supplies and electricity generation. On January 10, 2009 the Slovak government ordered the second reactor, then still undergoing a shutdown procedure, to be returned into power-generation-capable mode.[1] Eventually, the reactor was not reconnected to the grid and the final shutdown was resumed.[2]

Bohunice V2 (operating)

Bohunice V2 consits of two second generation VVER-440/213 and went online on August 20th 1984 and December 18th 1985, respectively. In November 2010 both reactors were upgraded from 440 MW to 505 MW and operation is planned to be maintained till 2025.

Bohunice V3 (planned)

In May 2009 Slovak Prime Minister Robert Fico and Czech PM Jan Fischer announced construction of a new reactor unit at Bohunice. A partnership of the Slovak Nuclear and Decommissioning Company (JAVYS) and the Czech energy company ČEZ will build the plant at a projected cost of around 3.7 billion Euros. The specification of new reactor has not been settled upon but will be rated at between 600 and 1,600 MWe. Four manufacturers have been mooted as possible suppliers: Mitsubishi, Atomenergoprojekt, Westinghouse and Areva. ČEZ is considering combining this project with their concurrent construction of the third and fourth reactors at Temelín Nuclear Power Station. Such a contract would be worth between 7 and 10 billion Euros.

Notes

External links

- Presentation about decommissioning of A1 (http://www.iync.org/archive/iync2004/presentation/track_b/b3_2/christofova_vasko.pdf#search=""A-1" bohunice")
- Electricite de France In Running To Build Slovak Nuclear Plant (http://www.energy-daily.com/reports/Electricite_de_France_In_Running_To_Build_Slovak_Nuclear_Plant_999.html)

Nuclear_reactor

This article is a subarticle of Nuclear power.

Core of CROCUS, a small nuclear reactor used for research at the EPFL in Switzerland

A **nuclear reactor** is a device to initiate and control a sustained nuclear chain reaction. Nuclear reactors are used at nuclear power plants for generating electricity and in propulsion of ships. Heat from nuclear fission is passed to a working fluid (water or gas), which runs through turbines. These either drive a ship's propellers or turn electrical generators. Nuclear generated steam in principle can be used for industrial process heat or for district heating. Some reactors are used to produce isotopes for medical and industrial use, or for production of plutonium for weapons. Some are run only for research.

Mechanism

Just as conventional power-stations generate electricity by harnessing the thermal energy released from burning fossil fuels, nuclear reactors convert the thermal energy released from nuclear fission.

Fission

When a large fissile atomic nucleus such as uranium-235 or plutonium-239 absorbs a neutron, it may undergo nuclear fission. The heavy nucleus splits into two or more lighter nuclei, (the fission products), releasing kinetic energy, gamma radiation, and free neutrons. A portion of these neutrons may later be absorbed by other fissile atoms and trigger further fission events, which release more neutrons, and so on. This is known as a nuclear chain reaction.

To control such a nuclear chain reaction, neutron poisons and neutron moderators can change the portion of neutrons that will go on to cause more fission.[] Nuclear reactors generally have automatic and manual systems to shut the fission reaction down if monitoring detects unsafe conditions.[1]

Commonly-used moderators include regular (light) water (in 74.8% of the world's reactors), solid graphite (20% of reactors) and heavy water (5% of reactors). Some experimental types of reactor have used beryllium, and hydrocarbons have been suggested as another possibility.[]

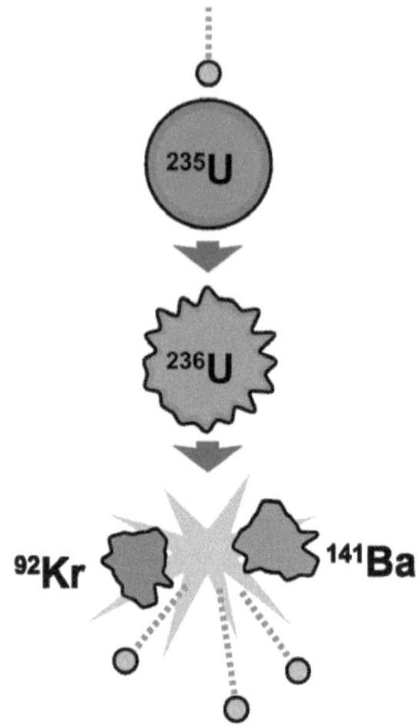

An induced nuclear fission event. A neutron is absorbed by the nucleus of a uranium-235 atom, which in turn splits into fast-moving lighter elements (fission products) and free neutrons. Though both reactors and nuclear weapons rely on nuclear chain-reactions, the rate of reactions in a reactor occurs much more slowly than in a bomb.

Heat generation

The reactor core generates heat in a number of ways:

- The kinetic energy of fission products is converted to thermal energy when these nuclei collide with nearby atoms.
- The reactor absorbs some of the gamma rays produced during fission and converts their energy converted into heat.
- Heat is produced by the radioactive decay of fission products and materials that have been activated by neutron absorption. This decay heat-source will remain for some time even after the reactor is shut down.

A kilogram of uranium-235 (U-235) converted via nuclear processes releases approximately three million times more energy than a kilogram of coal burned conventionally (7.2×10^{13} joules per kilogram of uranium-235 versus 2.4×10^{7} joules per kilogram of coal).[2] [3]

Cooling

A nuclear reactor coolant — usually water but sometimes a gas or a liquid metal (like liquid sodium) or molten salt — is circulated past the reactor core to absorb the heat that it generates. The heat is carried away from the reactor and is then used to generate steam. Most reactor systems employ a cooling system that is physically separated from the water that will be boiled to produce pressurized steam for the turbines, like the pressurized water reactor. But in some reactors the water for the steam turbines is boiled directly by the reactor core, for example the boiling water reactor.[]

Reactivity control

The power output of the reactor is adjusted by controlling how many neutrons are able to create more fissions.

Control rods that are made of a neutron poison are used to absorb neutrons. Absorbing more neutrons in a control rod means that there are fewer neutrons available to cause fission, so pushing the control rod deeper into the reactor will reduce its power output, and extracting the control rod will increase it.

At the first level of control in all nuclear reactors, a process of delayed neutron emission by a number of neutron-rich fission isotopes is an important physical process. These delayed neutrons account for about 0.65% of the total neutrons produced in fission, with the remainder (termed "prompt neutrons") released immediately upon fission. The fission products which produce delayed neutrons have half lives for their decay by neutron emission that range from milliseconds to as long as several minutes. Keeping the reactor in the zone of chain-reactivity where delayed neutrons are *necessary* to achieve a critical mass state, allows time for mechanical devices or human operators to have time to control a chain reaction in "real time"; otherwise the time between achievement of criticality and nuclear meltdown as a result of an exponential power surge from the normal nuclear chain reaction, would be too short to allow for intervention.

In some reactors, the coolant also acts as a neutron moderator. A moderator increases the power of the reactor by causing the fast neutrons that are released from fission to lose energy and become thermal neutrons. Thermal neutrons are more likely than fast neutrons to cause fission, so more neutron moderation means more power output from the reactors. If the coolant is a moderator, then temperature changes can affect the density of the coolant/moderator and therefore change power output. A higher temperature coolant would be less dense, and therefore a less effective moderator.

In other reactors the coolant acts as a poison by absorbing neutrons in the same way that the control rods do. In these reactors power output can be increased by heating the coolant, which makes it a less dense poison. Nuclear reactors generally have automatic and manual systems to scram the reactor in an emergency shut down. These systems insert large amounts of poison (often boron in the form of boric acid) into the reactor to shut the fission reaction down if unsafe conditions are detected or anticipated.[]

Most types of reactors are sensitive to a process variously known as xenon poisoning, or the iodine pit. Xenon-135 produced in the fission process acts as a "neutron poison" that absorbs neutrons and therefore tends to shut the reactor down. Xenon-135 accumulation can be controlled by keeping power levels high enough to destroy it as fast as it is produced. Fission also produces iodine-135, which in turn decays (with a half-life of under seven hours) to new xenon-135. When the reactor is shut down, iodine-135 continues to decay to xenon-135, making restarting the reactor more difficult for a day or two. This temporary state is the "iodine pit." If the reactor has sufficient extra reactivity capacity, it can be restarted. As the extra xenon-135 is transmuted to xenon-136 which is not a neutron poison, within a few hours the reactor experiences a "xenon burnoff (power) transient". Control rods must be further inserted to replace the neutron absorption of the lost xenon-135. Failure to properly follow such a procedure was a key step in the Chernobyl disaster.

Reactors used in nuclear marine propulsion (especially nuclear submarines) often cannot be run at continuous power around the clock in the same way that land-based power reactors are normally run, and in addition often need to have a very long core life without refueling. For this reason many designs use highly enriched uranium but incorporate burnable neutron poison directly into the fuel rods.[4] This allows the reactor to be constructed with a high excess of fissionable material, which is nevertheless made relatively more safe early in the reactor's fuel burn-cycle by the presence of the neutron-absorbing material which is later replaced by naturally produced long-lived neutron poisons (far longer-lived than xenon-135) which gradually accumulate over the fuel load's operating life.

Electrical power generation

The energy released in the fission process generates heat, some of which can be converted into usable energy. A common method of harnessing this thermal energy is to use it to boil water to produce pressurized steam which will then drive a steam turbine that turns an alternator and generates electricity.[]

Early reactors

The neutron was discovered in 1932. The concept of a nuclear chain reaction brought about by nuclear reactions mediated by neutrons, was first realized shortly thereafter, by Hungarian scientist Leó Szilárd, in 1933. He filed a patent for his idea of a simple nuclear reactor the following year while working at the Admiralty in London.[5] However, Szilárd's idea did not incorporate the idea of nuclear fission as a neutron source, since that process was not yet discovered. Szilárd's ideas for nuclear reactors using neutron-mediated nuclear chain reactions in light elements proved unworkable.

Inspiration for a new type of reactor using uranium came from the discovery by Lise Meitner, Fritz Strassmann and Otto Hahn in 1938 that bombardment of uranium with neutrons (provided by an alpha-on-beryllium fusion reaction, a "neutron howitzer") produced a barium residue, which they reasoned was created by the fissioning of the uranium nuclei. Subsequent studies in early 1939 (one of them by Szilárd and Fermi) revealed that several neutrons were also released during the fissioning, making available the opportunity for the nuclear chain reaction that Szilárd had envisioned six years previously.

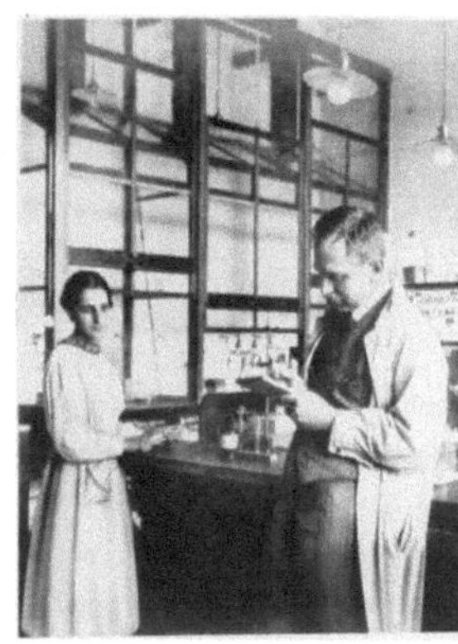

Lise Meitner and Otto Hahn in their laboratory.

On August 2, 1939 Albert Einstein signed a letter to President Franklin D. Roosevelt (written by Szilárd) suggesting that the discovery of uranium's fission could lead to the development of "extremely powerful bombs of a new type", giving impetus to the study of reactors and fission. Szilárd and Einstein knew each other well and had worked together years previously, but Einstein had never thought about this possibility for nuclear energy until Szilard reported it to him, at the beginning of his quest to produce the Einstein-Szilárd letter to alert the U.S. government.

Shortly after, Hitler's Germany invaded Poland in 1939, starting World War II in Europe. The U.S. was not yet officially at war, but in October, when the Einstein-Szilárd letter was delivered to Roosevelt, he commented that the purpose of doing the research was to make sure "the Nazis don't blow us up." The U.S. nuclear project followed, although with some delay as there remained skepticism (some of it from Fermi) and also little action from the small number of officials in the government who were initially charged with moving the project forward.

The following year the U.S. Government received the Frisch–Peierls memorandum from the UK, which stated that the amount of uranium needed for a chain reaction was far lower than had previously been thought. The memorandum was a product of the MAUD Committee, which was working on the UK atomic bomb project, known as Tube Alloys, later to be subsumed within the Manhattan Project.

Eventually, the first artificial nuclear reactor, Chicago Pile-1, was constructed at the University of Chicago, by a team led by Enrico Fermi, in late 1942. By this time, the program had been pressured for a year by U.S. entry into the war. The Chicago Pile achieved criticality on December 2, 1942[6] at 3:25 PM. The reactor support structure was made of wood, which supported a pile (hence the name) of graphite blocks, embedded in which was natural uranium-oxide 'pseudospheres' or 'briquettes'.

The Chicago Pile Team, including Enrico Fermi and Leó Szilárd.

Soon after the Chicago Pile, the U.S. military developed a number of nuclear reactors for the Manhattan Project starting in 1943. The primary purpose for the largest reactors (located at the Hanford Site in Washington state), was the mass production of plutonium for nuclear weapons. Fermi and Szilard applied for a patent on reactors on 19 December 1944. Its issuance was delayed for 10 years because of wartime secrecy.[7]

"World's first nuclear power plant" is the claim made by signs at the site of the EBR-I, which is now a museum near Arco, Idaho. This experimental LMFBR operated by the U.S. Atomic Energy Commission produced 0.8 kW in a test on December 20, 1951[8] and 100 kW (electrical) the following day,[9] having a design output of 200 kW (electrical).

Besides the military uses of nuclear reactors, there were political reasons to pursue civilian use of atomic energy. U.S. President Dwight Eisenhower made his famous Atoms for Peace speech to the UN General Assembly on December 8, 1953. This diplomacy led to the dissemination of reactor technology to U.S. institutions and worldwide.

The first nuclear power plant built for civil purposes was the AM-1 Obninsk Nuclear Power Plant, launched on June 27, 1954 in the Soviet Union. It produced around 5 MW (electrical).

After World War II, the U.S. military sought other uses for nuclear reactor technology. Research by the Army and the Air Force never came to fruition; however, the U.S. Navy succeeded when they steamed the USS *Nautilus* (SSN-571) on nuclear power January 17, 1955.

The first commercial nuclear power station, Calder Hall in Sellafield, England was opened in 1956 with an initial capacity of 50 MW (later 200 MW).[] []

The first portable nuclear reactor "Alco PM-2A" used to generate electrical power (2 MW) for Camp Century from 1960.[10]

Components

The key components common to most types of nuclear power plants are:

The control room of NC State's Pulstar Nuclear Reactor.

- Nuclear fuel
- Nuclear reactor core
- Neutron moderator
- Neutron poison
- Neutron howitzer (provides steady source of neutrons to re-initiate reaction following shutdown)
- Coolant (often the Neutron Moderator and the Coolant are the same, usually both purified water)
- Control rods
- Reactor vessel
- Boiler feedwater pump
- Steam generators (not in BWRs)
- Steam turbine
- Electrical generator
- Condenser
- Cooling tower (not always required)
- Radwaste System (a section of the plant handling radioactive waste)
- Refueling Floor
- Spent fuel pool
- Nuclear safety systems
  - Reactor Protective System (RPS)
  - Emergency Diesel Generators
  - Emergency Core Cooling Systems (ECCS)
  - Standby Liquid Control System (emergency boron injection, in BWRs only)
- Essential service water system (ESWS)
- Containment building
- Control room
- Emergency Operations Facility
- Nuclear training facility (usually contains a Control Room simulator)

Reactor types

Classifications

Nuclear Reactors are classified by several methods; a brief outline of these classification methods is provided.

NC State's PULSTAR Reactor is a 1 MW pool-type research reactor with 4% enriched, pin-type fuel consisting of UO_2 pellets in zircaloy cladding.

Classification by type of nuclear reaction

- Nuclear fission. All commercial power reactors are based on nuclear fission. They generally use uranium and its product plutonium as nuclear fuel, though a thorium fuel cycle is also possible. Fission reactors can be divided roughly into two classes, depending on the energy of the neutrons that sustain the fission chain reaction:
  - Thermal reactors use slowed or thermal neutrons to keep up the fission of their fuel. Almost all current reactors are of this type. These contain neutron moderator materials that slow neutrons until their neutron temperature is *thermalized*, that is, until their kinetic energy approaches the average kinetic energy of the surrounding particles. Thermal neutrons have a far higher cross section (probability) of fissioning the fissile nuclei uranium-235, plutonium-239, and plutonium-241, and a relatively lower probability of neutron capture by uranium-238 (U-238) compared to the faster neutrons that originally result from fission, allowing use of low-enriched uranium or even natural uranium fuel. The moderator is often also the coolant, usually water under high pressure to increase the boiling point. These are surrounded by a reactor vessel, instrumentation to monitor and control the reactor, radiation shielding, and a containment building.
  - Fast neutron reactors use fast neutrons to cause fission in their fuel. They do not have a neutron moderator, and use less-moderating coolants. Maintaining a chain reaction requires the fuel to be more highly enriched in fissile material (about 20% or more) due to the relatively lower probability of fission versus capture by U-238. Fast reactors have the potential to produce less transuranic waste because all actinides are fissionable with fast neutrons,[11] but they are more difficult to build and more expensive to operate. Overall, fast reactors are less common than thermal reactors in most applications. Some early power stations were fast reactors, as are some Russian naval propulsion units. Construction of prototypes is continuing (see fast breeder or generation IV reactors).
- Nuclear fusion. Fusion power is an experimental technology, generally with hydrogen as fuel. While not suitable for power production, Farnsworth-Hirsch fusors are used to produce neutron radiation.

Classification by moderator material

Used by thermal reactors:

- Graphite moderated reactors
- Water moderated reactors
  - Heavy water reactors
  - Light water moderated reactors (LWRs). Light water reactors use ordinary water to moderate and cool the reactors. When at operating temperature, if the temperature of the water increases, its density drops, and fewer neutrons passing through it are slowed enough to trigger further reactions. That negative feedback stabilizes the reaction rate. Graphite and heavy water reactors tend to be more thoroughly thermalised than light water reactors. Due to the extra thermalization, these types can use natural uranium/unenriched fuel.

- Light element moderated reactors. These reactors are moderated by lithium or beryllium.
  - Molten salt reactors (MSRs) are moderated by a light elements such as lithium or beryllium, which are constituents of the coolant/fuel matrix salts LiF and BeF_2.
  - Liquid metal cooled reactors, such as one whose coolant is a mixture of Lead and Bismuth, may use BeO as a moderator.
- Organically moderated reactors (OMR) use biphenyl and terphenyl as moderator and coolant.

Classification by coolant

Treatment of the interior part of a VVER-1000 reactor frame on Atommash.

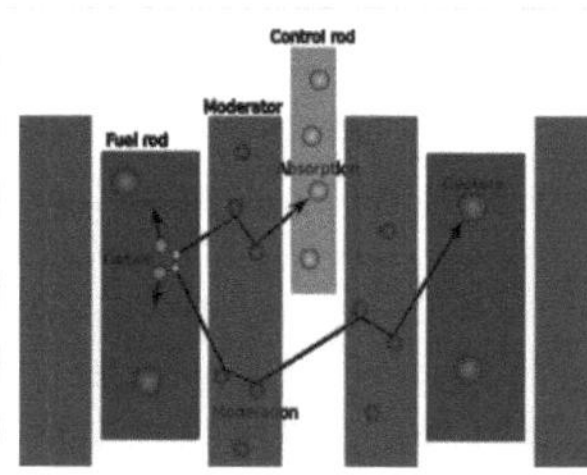

In thermal nuclear reactors (LWRs in specific), the coolant acts as a moderator that must slow down the neutrons before they can be efficiently absorbed by the fuel.

- Water cooled reactor. There are 104 operating reactors in the United States. Of these, 69 are pressurized water reactors (PWR), and 35 are boiling water reactors (BWR).[12]
  - Pressurized water reactor (PWR)
    - A primary characteristic of PWRs is a pressurizer, a specialized pressure vessel. Most commercial PWRs and naval reactors use pressurizers. During normal operation, a pressurizer is partially filled with water, and a steam bubble is maintained above it by heating the water with submerged heaters. During normal operation, the pressurizer is connected to the primary reactor pressure vessel (RPV) and the pressurizer "bubble" provides an expansion space for changes in water volume in the reactor. This arrangement also provides a means of pressure control for the reactor by increasing or decreasing the steam pressure in the pressurizer using the pressurizer heaters.
    - Pressurised heavy water reactors are a subset of pressurized water reactors, sharing the use of a pressurized, isolated heat transport loop, but using heavy water as coolant and moderator for the greater neutron economies it offers.
  - Boiling water reactor (BWR)
    - BWRs are characterized by boiling water around the fuel rods in the lower portion of a primary reactor pressure vessel. A boiling water reactor uses ^{235}U, enriched as uranium dioxide, as its fuel. The fuel is assembled into rods that are submerged in water and housed in a steel vessel. The nuclear fission causes the water to boil, generating steam. This steam flows through pipes into turbines. The turbines are driven by the steam, and this process generates electricity.[] During normal operation, pressure is controlled by the amount of steam flowing from the reactor pressure vessel to the turbine.
  - Pool-type reactor
- Liquid metal cooled reactor. Since water is a moderator, it cannot be used as a coolant in a fast reactor. Liquid metal coolants have included sodium, NaK, lead, lead-bismuth eutectic, and in early reactors, mercury.
  - Sodium-cooled fast reactor
  - Lead-cooled fast reactor
- Gas cooled reactors are cooled by a circulating inert gas, often helium in high-temperature designs, while carbon dioxide has been used in past British and French nuclear power plants. Nitrogen has also been used. Utilization of the heat varies, depending on the reactor. Some reactors run hot enough that the gas can directly power a gas

turbine. Older designs usually run the gas through a heat exchanger to make steam for a steam turbine.

- Molten Salt Reactors (MSRs) are cooled by circulating a molten salt, typically a eutectic mixture of fluoride salts, such as FLiBe. In a typical MSR, the coolant is also used as a matrix in which the fissile material is dissolved.

Classification by generation

- Generation I reactor (early prototypes, research reactors, non-commercial power producing reactors)
- Generation II reactor (most current nuclear power plants 1965-1996)
- Generation III reactor (evolutionary improvements of existing designs 1996-now)
- Generation IV reactor (technologies still under development unknown start date, possibly 2030)

The "Gen IV"-term was dubbed by the United States Department of Energy (DOE) for developing new plant types in 2000.[13] In 2003, the French Commissariat à l'Énergie Atomique (CEA) was the first to refer to Gen II types in Nucleonics Week; .[14] First mentioning of Gen III was also in 2000 in conjunction with the launch of the Generation IV International Forum (GIF) plans.

Classification by phase of fuel

- Solid fueled
- Fluid fueled
  - Aqueous homogeneous reactor
  - Molten salt reactor
- Gas fueled (theoretical)

Classification by use

- Electricity
  - Nuclear power plants including small modular reactors
- Propulsion, see nuclear propulsion
  - Nuclear marine propulsion
  - Various proposed forms of rocket propulsion
- Other uses of heat
  - Desalination
  - Heat for domestic and industrial heating
  - Hydrogen production for use in a hydrogen economy
- Production reactors for transmutation of elements
  - Breeder reactors are capable of producing more fissile material than they consume during the fission chain reaction (by converting fertile U-238 to Pu-239, or Th-232 to U-233). Thus, a uranium breeder reactor, once running, can be re-fueled with natural or even depleted uranium, and a thorium breeder reactor can be re-fueled with thorium; however, an initial stock of fissile material is required.[15]
  - Creating various radioactive isotopes, such as americium for use in smoke detectors, and cobalt-60, molybdenum-99 and others, used for imaging and medical treatment.
  - Production of materials for nuclear weapons such as weapons-grade plutonium
- Providing a source of neutron radiation (for example with the pulsed Godiva device) and positron radiation (e.g. neutron activation analysis and potassium-argon dating)
- Research reactor: Typically reactors used for research and training, materials testing, or the production of radioisotopes for medicine and industry. These are much smaller than power reactors or those propelling ships, and many are on university campuses. There are about 280 such reactors operating, in 56 countries. Some operate with high-enriched uranium fuel, and international efforts are underway to substitute low-enriched fuel.[16]

Current technologies

Pressurized Water Reactors (PWR)

Diablo Canyon — a PWR

These reactors use a pressure vessel to contain the nuclear fuel, control rods, moderator, and coolant. They are cooled and moderated by high pressure liquid water. The hot radioactive water that leaves the pressure vessel is looped through a steam generator, which in turn heats a secondary (non-radioactive) loop of water to steam that can run turbines. They are the majority of current reactors. This is a thermal neutron reactor design, the newest of which are the VVER-1200, Advanced Pressurized Water Reactor and the European Pressurized Reactor. United States Naval reactors are of this type.

Boiling Water Reactors (BWR)

Laguna Verde nuclear power plant — a BWR

A BWR is like a PWR without the steam generator. A boiling water reactor is cooled and moderated by water like a PWR, but at a lower pressure, which allows the water to boil inside the pressure vessel producing the steam that runs the turbines. Unlike a PWR, there is no primary and secondary loop. The thermal efficiency of these reactors can be higher, and they can be simpler, and even potentially more stable and safe. This is a thermal neutron reactor design, the newest of which are the Advanced Boiling Water Reactor and the Economic Simplified Boiling Water Reactor.

Pressurised heavy water reactorPressurized Heavy Water Reactor (PHWR)

The CANDU Qinshan Nuclear Power Plant

A Canadian design (known as CANDU), these reactors are heavy-water-cooled and -moderated Pressurized-Water reactors. Instead of using a single large pressure vessel as in a PWR, the fuel is contained in hundreds of pressure tubes. These reactors are fueled with natural uranium and are thermal neutron reactor designs. PHWRs can be refueled while at full power, which makes them very efficient in their use of uranium (it allows for precise flux control in the core). CANDU PHWRs have been built in Canada, Argentina, China, India, Pakistan, Romania, and South Korea. India also operates a number of PHWRs, often termed 'CANDU-derivatives', built after the Government of Canada halted nuclear dealings with India following the 1974 Smiling Buddha nuclear weapon test.

Reaktor Bolshoy Moschnosti Kanalniy (High Power Channel Reactor) (RBMK)

The Ignalina Nuclear Power Plant — a RBMK type (closed 2009)

A Soviet design, built to produce plutonium as well as power. RBMKs are water cooled with a graphite moderator. RBMKs are in some respects similar to CANDU in that they are refuelable during power operation and employ a pressure tube design instead of a PWR-style pressure vessel. However, unlike CANDU they are very unstable and large, making containment buildings for them expensive. A series of critical safety flaws have also been identified with the RBMK design, though some of these were corrected following the Chernobyl disaster. Their main attraction is their use of light water and un-enriched uranium. As of 2010, 11 remain open, mostly due to safety improvements and help from international safety agencies such as the DOE. Despite these safety improvements, RBMK reactors are still considered one of the most dangerous reactor designs in use. RBMK reactors were deployed only in the former Soviet Union.

Gas Cooled Reactor (GCR) and Advanced gas cooled reactorAdvanced Gas Cooled Reactor (AGR)

The Magnox Sizewell A nuclear power station

The Torness nuclear power station — an AGR

These are generally graphite moderated and CO_2 cooled. They can have a high thermal efficiency compared with PWRs due to higher operating temperatures. There are a number of operating reactors of this design, mostly in the United Kingdom, where the concept was developed. Older designs (i.e. Magnox stations) are either shut down or will be in the near future. However, the AGCRs have an anticipated life of a further 10 to 20 years. This is a thermal neutron reactor design. Decommissioning costs can be high due to large volume of reactor core.

Breeder reactorLiquid Metal Fast breeder reactorFast Breeder Reactor (LMFBR)

This is a reactor design that is cooled by liquid metal, totally unmoderated, and produces more fuel than it consumes. They are said to "breed" fuel, because they produce fissionable fuel during operation because of neutron capture. These reactors can function much like a PWR in terms of efficiency, and do not require much high pressure containment, as the liquid metal does not need to be kept at high pressure, even at very high temperatures. BN-350 and BN-600 in USSR and Superphénix in France were a reactor of this type, as was Fermi-I in the United States. The Monju reactor in Japan suffered a sodium leak in 1995 and was restarted in May 2010. All of them use/used liquid sodium. These reactors are fast neutron, not thermal neutron designs. These reactors come in two types:

Lead cooled

The Superphénix, one of the few FBRs

Using lead as the liquid metal provides excellent radiation shielding, and allows for operation at very high temperatures. Also, lead is (mostly) transparent to neutrons, so fewer neutrons are lost in the coolant, and the coolant does not become radioactive. Unlike sodium, lead is mostly inert, so there is less risk of explosion or accident, but such large quantities of lead may be problematic from toxicology and disposal points of view. Often a reactor of this type would use a lead-bismuth eutectic mixture. In this case, the bismuth would present some minor radiation problems, as it is not quite as transparent to neutrons, and can be transmuted to a radioactive isotope more readily than lead. The Russian Alfa class submarine uses a lead-bismuth-cooled fast reactor as its main power plant.

Sodium cooled

Most LMFBRs are of this type. The sodium is relatively easy to obtain and work with, and it also manages to actually prevent corrosion on the various reactor parts immersed in it. However, sodium explodes violently when exposed to water, so care must be taken, but such explosions would not be vastly more violent than (for example) a leak of superheated fluid from a SCWR or PWR. EBR-I, the first reactor to have a core meltdown, was of this type.

Pebble Bed Reactors (PBR)

These use fuel molded into ceramic balls, and then circulate gas through the balls. The result is an efficient, low-maintenance, very safe reactor with inexpensive, standardized fuel. The prototype was the AVR.

Molten Salt Reactors

These dissolve the fuels in fluoride salts, or use fluoride salts for coolant. These have many safety features, high efficiency and a high power density suitable for vehicles. Notably, they have no high pressures or flammable components in the core. The prototype was the MSRE, which also used Thorium's fuel cycle to produce 0.1% of the radioactive waste of standard reactors.

Aqueous Homogeneous Reactor (AHR)

These reactors use soluble nuclear salts dissolved in water and mixed with a coolant and a neutron moderator.

Future and developing technologies

Advanced reactors

More than a dozen advanced reactor designs are in various stages of development.[] Some are evolutionary from the PWR, BWR and PHWR designs above, some are more radical departures. The former include the Advanced Boiling Water Reactor (ABWR), two of which are now operating with others under construction, and the planned passively safe Economic Simplified Boiling Water Reactor (ESBWR) and AP1000 units (see Nuclear Power 2010 Program).

- The Integral Fast Reactor (IFR) was built, tested and evaluated during the 1980s and then retired under the Clinton administration in the 1990s due to nuclear non-proliferation policies of the administration. Recycling spent fuel is the core of its design and it therefore produces only a fraction of the waste of current reactors.[]
- The Pebble Bed Reactor, a High Temperature Gas Cooled Reactor (HTGCR), is designed so high temperatures reduce power output by Doppler broadening of the fuel's neutron cross-section. It uses ceramic fuels so its safe operating temperatures exceed the power-reduction temperature range. Most designs are cooled by inert helium. Helium is not subject to steam explosions, resists neutron absorption leading to radioactivity, and does not dissolve contaminants that can become radioactive. Typical designs have more layers (up to 7) of passive containment than light water reactors (usually 3). A unique feature that may aid safety is that the fuel-balls actually form the core's mechanism, and are replaced one-by-one as they age. The design of the fuel makes fuel reprocessing expensive.
- The Small, sealed, transportable, autonomous reactor (SSTAR) is being primarily researched and developed in the US, intended as a fast breeder reactor that is passively safe and could be remotely shut down in case the suspicion arises that it is being tampered with.
- The Clean And Environmentally Safe Advanced Reactor (CAESAR) is a nuclear reactor concept that uses steam as a moderator — this design is still in development.
- The Hydrogen Moderated Self-regulating Nuclear Power Module (HPM) is a reactor design emanating from the Los Alamos National Laboratory that uses uranium hydride as fuel.
- Subcritical reactors are designed to be safer and more stable, but pose a number of engineering and economic difficulties. One example is the Energy amplifier.
- Thorium based reactors. It is possible to convert Thorium-232 into U-233 in reactors specially designed for the purpose. In this way, thorium, which is more plentiful than uranium, can be used to breed U-233 nuclear fuel. U-233 is also believed to have favourable nuclear properties as compared to traditionally used U-235, including better neutron economy and lower production of long lived transuranic waste.
    - Advanced Heavy Water Reactor (AHWR)— A proposed heavy water moderated nuclear power reactor that will be the next generation design of the PHWR type. Under development in the Bhabha Atomic Research Centre (BARC), India.
    - KAMINI — A unique reactor using Uranium-233 isotope for fuel. Built in India by BARC and Indira Gandhi Center for Atomic Research (IGCAR).
    - India is also planning to build fast breeder reactors using the thorium – Uranium-233 fuel cycle. The FBTR (Fast Breeder Test Reactor) in operation at Kalpakkam (India) uses Plutonium as a fuel and liquid sodium as a coolant.

Generation IV reactors

Generation IV reactors are a set of theoretical nuclear reactor designs currently being researched. These designs are generally not expected to be available for commercial construction before 2030. Current reactors in operation around the world are generally considered second- or third-generation systems, with the first-generation systems having been retired some time ago. Research into these reactor types was officially started by the Generation IV International Forum (GIF) based on eight technology goals. The primary goals being to improve nuclear safety, improve proliferation resistance, minimize waste and natural resource utilization, and to decrease the cost to build and run such plants.[]

- Gas cooled fast reactor
- Lead cooled fast reactor
- Molten salt reactor
- Sodium-cooled fast reactor
- Supercritical water reactor
- Very high temperature reactor

Generation V+ reactors

Generation V reactors are designs which are theoretically possible, but which are not being actively considered or researched at present. Though such reactors could be built with current or near term technology, they trigger little interest for reasons of economics, practicality, or safety.

- Liquid core reactor. A closed loop liquid core nuclear reactor, where the fissile material is molten uranium or uranium solution cooled by a working gas pumped in through holes in the base of the containment vessel.
- Gas core reactor. A closed loop version of the nuclear lightbulb rocket, where the fissile material is gaseous uranium-hexafluoride contained in a fused silica vessel. A working gas (such as hydrogen) would flow around this vessel and absorb the UV light produced by the reaction. In theory, using UF_6 as a working fuel directly (rather than as a stage to one, as is done now) would mean lower processing costs, and very small reactors. In practice, running a reactor at such high power densities would probably produce unmanageable neutron flux.
  - Gas core EM reactor. As in the Gas Core reactor, but with photovoltaic arrays converting the UV light directly to electricity.
- Fission fragment reactor

Fusion reactors

Controlled nuclear fusion could in principle be used in fusion power plants to produce power without the complexities of handling actinides, but significant scientific and technical obstacles remain. Several fusion reactors have been built, but as yet none has 'produced' more thermal energy than electrical energy consumed. Despite research having started in the 1950s, no commercial fusion reactor is expected before 2050. The ITER project is currently leading the effort to commercialize fusion power.

Nuclear fuel cycle

Thermal reactors generally depend on refined and enriched uranium. Some nuclear reactors can operate with a mixture of plutonium and uranium (see MOX). The process by which uranium ore is mined, processed, enriched, used, possibly reprocessed and disposed of is known as the nuclear fuel cycle.

Under 1% of the uranium found in nature is the easily fissionable U-235 isotope and as a result most reactor designs require enriched fuel. Enrichment involves increasing the percentage of U-235 and is usually done by means of gaseous diffusion or gas centrifuge. The enriched result is then converted into uranium dioxide powder, which is pressed and fired into pellet form. These pellets are stacked into tubes which are then sealed and called fuel rods. Many of these fuel rods are used in each nuclear reactor.

Most BWR and PWR commercial reactors use uranium enriched to about 4% U-235, and some commercial reactors with a high neutron economy do not require the fuel to be enriched at all (that is, they can use natural uranium). According to the International Atomic Energy Agency there are at least 100 research reactors in the world fueled by highly enriched (weapons-grade/90% enrichment uranium). Theft risk of this fuel (potentially used in the production of a nuclear weapon) has led to campaigns advocating conversion of this type of reactor to low-enrichment uranium (which poses less threat of proliferation).[17]

Fissile U-235 and non-fissile but fissionable and fertile U-238 are both used in the fission process. U-235 is fissionable by thermal (i.e. slow-moving) neutrons. A thermal neutron is one which is moving about the same speed as the atoms around it. Since all atoms vibrate proportionally to their absolute temperature, a thermal neutron has the best opportunity to fission U-235 when it is moving at this same vibrational speed. On the other hand, U-238 is more likely to capture a neutron when the neutron is moving very fast. This U-239 atom will soon decay into plutonium-239, which is another fuel. Pu-239 is a viable fuel and must be accounted for even when a highly enriched uranium fuel is used. Plutonium fissions will dominate the U-235 fissions in some reactors, especially after the initial loading of U-235 is spent. Plutonium is fissionable with both fast and thermal neutrons, which make it ideal for either nuclear reactors or nuclear bombs.

Most reactor designs in existence are thermal reactors and typically use water as a neutron moderator (moderator means that it slows down the neutron to a thermal speed) and as a coolant. But in a fast breeder reactor, some other kind of coolant is used which will not moderate or slow the neutrons down much. This enables fast neutrons to dominate, which can effectively be used to constantly replenish the fuel supply. By merely placing cheap unenriched uranium into such a core, the non-fissionable U-238 will be turned into Pu-239, "breeding" fuel.

In thorium fuel cycle thorium-232 absorbs a neutron in either a fast or thermal reactor. The thorium-233 beta decays to protactinium-233 and then to uranium-233, which in turn is used as fuel. Hence, like uranium-238, thorium-232 is a fertile material.

Fueling of nuclear reactors

The amount of energy in the reservoir of nuclear fuel is frequently expressed in terms of "full-power days," which is the number of 24-hour periods (days) a reactor is scheduled for operation at full power output for the generation of heat energy. The number of full-power days in a reactor's operating cycle (between refueling outage times) is related to the amount of fissile uranium-235 (U-235) contained in the fuel assemblies at the beginning of the cycle. A higher percentage of U-235 in the core at the beginning of a cycle will permit the reactor to be run for a greater number of full-power days.

At the end of the operating cycle, the fuel in some of the assemblies is "spent" and is discharged and replaced with new (fresh) fuel assemblies, although in practice it is the buildup of reaction poisons in nuclear fuel that determines the lifetime of nuclear fuel in a reactor. Long before all possible fission has taken place, the buildup of long-lived neutron absorbing fission byproducts impedes the chain reaction. The fraction of the reactor's fuel core replaced during refueling is typically one-fourth for a boiling-water reactor and one-third for a pressurized-water reactor. The disposition and storage of this spent fuel is one of the most challenging aspects of the operation of a commercial nuclear power plant. This nuclear waste is highly radioactive and its toxicity presents a danger for thousands of years.[]

Not all reactors need to be shut down for refueling; for example, pebble bed reactors, RBMK reactors, molten salt reactors, Magnox, AGR and CANDU reactors allow fuel to be shifted through the reactor while it is running. In a CANDU reactor, this also allows individual fuel elements to be situated within the reactor core that are best suited to the amount of U-235 in the fuel element.

The amount of energy extracted from nuclear fuel is called its burnup, which is expressed in terms of the heat energy produced per initial unit of fuel weight. Burn up is commonly expressed as megawatt days thermal per metric ton of initial heavy metal.

Safety

Nuclear safety covers the actions taken to prevent nuclear and radiation accidents or to limit their consequences. The nuclear power industry has improved the safety and performance of reactors, and has proposed new safer (but generally untested) reactor designs but there is no guarantee that the reactors will be designed, built and operated correctly.[] Mistakes do occur and the designers of reactors at Fukushima in Japan did not anticipate that a tsunami generated by an earthquake would disable the backup systems that were supposed to stabilize the reactor after the earthquake.[18] According to UBS AG, the Fukushima I nuclear accidents have cast doubt on whether even an advanced economy like Japan can master nuclear safety.[19] Catastrophic scenarios involving terrorist attacks are also conceivable.[] An interdisciplinary team from MIT have estimated that given the expected growth of nuclear power from 2005–2055, at least four serious nuclear accidents would be expected in that period.[20]

Accidents

Some serious nuclear and radiation accidents have occurred. Nuclear power plant accidents include the Chernobyl disaster (1986), Fukushima Daiichi nuclear disaster (2011), the Three Mile Island accident (1979), and SL-1 accident (1961).[22] Nuclear-powered submarine mishaps include the K-19 reactor accident (1961),[23] the K-27 reactor accident (1968),[] and the K-431 reactor accident (1985).[22]

Three of the reactors at Fukushima I overheated, causing meltdowns that eventually led to explosions, which released large amounts of radioactive material into the air.[21]

Nuclear reactors have been launched into Earth orbit at least 34 times. A number of incidents connected with the unmanned nuclear-reactor-powered Soviet RORSAT radar satellite program resulted in spent nuclear fuel re-entering the Earth's atmosphere from orbit.

Natural nuclear reactors

Although nuclear fission reactors are often thought of as being solely a product of modern technology, the first nuclear fission reactors were in fact naturally occurring. A natural nuclear fission reactor can occur under certain circumstances that mimic the conditions in a constructed reactor.[24] Fifteen natural fission reactors have so far been found in three separate ore deposits at the Oklo mine in Gabon, West Africa. First discovered in 1972 by French physicist Francis Perrin, they are collectively known as the Oklo Fossil Reactors. Self-sustaining nuclear fission reactions took place in these reactors approximately 1.5 billion years ago, and ran for a few hundred thousand years, averaging 100 kW of power output during that time.[25] The concept of a natural nuclear reactor was theorized as early as 1956 by Paul Kuroda at the University of Arkansas.[26] []

Such reactors can no longer form on Earth: radioactive decay over this immense time span has reduced the proportion of U-235 in naturally occurring uranium to below the amount required to sustain a chain reaction.

The natural nuclear reactors formed when a uranium-rich mineral deposit became inundated with groundwater that acted as a neutron moderator, and a strong chain reaction took place. The water moderator would boil away as the reaction increased, slowing it back down again and preventing a meltdown. The fission reaction was sustained for hundreds of thousands of years.

These natural reactors are extensively studied by scientists interested in geologic radioactive waste disposal. They offer a case study of how radioactive isotopes migrate through the Earth's crust. This is a significant area of controversy as opponents of geologic waste disposal fear that isotopes from stored waste could end up in water supplies or be carried into the environment.

See also

- List of nuclear reactors
- List of United States Naval reactors
- List of small nuclear reactor designs
- Nuclear marine propulsion
- Neutron transport
- Nuclear power by country
- Radioisotope thermoelectric generator
- Safety engineering
- Small modular reactor
- Thorium-based nuclear power
- World Nuclear Industry Status Report

References

[5] L. Szilárd, "Improvements in or relating to the transmutation of chemical elements," (http://v3.espacenet.com/textdoc?DB=EPODOC&IDX=GB630726) British patent number: GB630726 (filed: 28 June 1934; published: 30 March 1936).

[6] The First Reactor, U.S. Atomic Energy Commission, Division of Technical Information

[7] U.S. Patent 2708656 (http://www.google.com/patents/US2708656) "Neutronic Reactor " issued 17 May 1955

[8] Experimental Breeder Reactor 1 factsheet (https://inlportal.inl.gov/portal/server.pt/gateway/PTARGS_0_200_816_259_0_43/http;/inlpublisher;7087/publishedcontent/publish/communities/inl_gov/about_inl/home_page_fact_sheets/sheets/experimental_breeder_reactor___i_4.pdf), Idaho National Laboratory

[14] Nucleonics Week, Vol. 44, No. 39; Pg. 7, 25 September 2003 Quote: "Etienne Pochon, CEA director of nuclear industry support, outlined EPR's improved performance and enhanced safety features compared to the advanced Generation II designs on which it was based."

[15] A Technology Roadmap for Generation IV Nuclear Energy Systems (http://www.gen-4.org/PDFs/GenIVRoadmap.pdf) PDF (4.33 MB); see "Fuel Cycles and Sustainability"

[22] The Worst Nuclear Disasters (http://www.time.com/time/photogallery/0,29307,1887705,00.html)

[23] Strengthening the Safety of Radiation Sources (http://www.iaea.org/Publications/Magazines/Bulletin/Bull413/article1.pdf) p. 14.

[24] Video of physics lecture (http://video.google.com/videoplay?docid=-2334857802602777622) – at Google Video; a natural nuclear reactor is mentioned at 42:40 mins into the video

[25] Meshik, Alex P. "The Workings of an Ancient Nuclear Reactor." *Scientific American.* November, 2005. Pg. 82.

External links

- Uranium Conference adds discussion of Japan accident (http://www.nmt.edu/nmt-news/336-2011/4101-uranium-conference-adds-session-reviewing-implications-of-japan-accident)
- Boiling Water Reactor Plant Technology Education (http://www.acme-nuclear.com) — Includes the PC-based BWR reactor simulation.
- World Nuclear Association — How it Works (http://www.world-nuclear.org/how/how.html)
- A Debate: Is Nuclear Power The Solution to Global Warming? (http://www.democracynow.org/article.pl?sid=04/09/24/1359225)
- Union of Concerned Scientists, Concerns re: US nuclear reactor program (http://www.ucsusa.org/clean_energy/nuclear_safety/page.cfm?pageID=1408)
- Freeview Video 'Nuclear Power Plants — What's the Problem' A Royal Institution Lecture by John Collier by the Vega Science Trust. (http://www.vega.org.uk/video/programme/67)
- U.S. plants and operators (http://www.nucleartourist.com/basics/current.htm)
- Glossary of Nuclear Terms (http://www.unionmillwright.com/nuke.html)
- American Nuclear Society — Glossary of Terms (http://www.ans.org/pi/resources/glossary/)
- Nuclear Energy Institute — How it Works: Electric Power Generation (http://www.nei.org/howitworks/electricpowergeneration/)

- Annotated bibliography of nuclear reactor technology from the Alsos Digital Library (http://alsos.wlu.edu/qsearch.aspx?browse=science/Nuclear+Reactors)

Jaslovské_Bohunice

Jaslovské Bohunice	
Village	
Country	Slovakia
Region	Trnava
District	Trnava
Elevation	160 m (525 ft)
Coordinates	
Area	20.083 km^2 (8 sq mi)
Population	1893 (*2006-12-31*)
Density	94 / km^2 (243 / sq mi)
First mentioned	1113
Timezone	CET (UTC+1)
- summer (DST)	CEST (UTC+2)
Postal code	919 30
Area code	+421-33
Car plate	TT
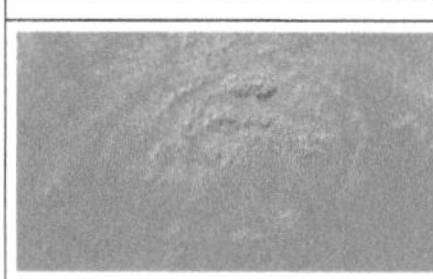 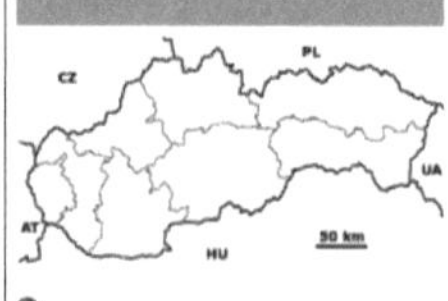 Location in Slovakia	

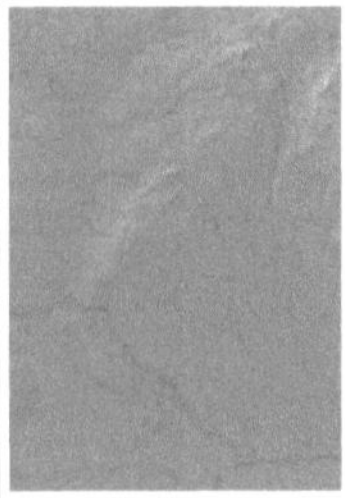

Location in the Trnava Region

Statistics: MOŠ/MIS [1]
Website: jaslovskebohunice.sk [2]

Jaslovské Bohunice is a small village in Slovakia in the Trnava District.[3] It is best known for the nearby Bohunice Nuclear Power Plants complex.

The village arose in 1958 through a merge of Jaslovce (pop. 578 in 1948) and Bohunice (pop. 619 in 1948). The former village of Paderovce was amalgamated in 1976. The first written reference dates back to 1113 (to Bohunice).

References

[1] http://www.statistics.sk/mosmis/eng/prvav2.jsp?txtUroven=420207&lstObec=507156&Okruh=zaklad
[2] http://jaslovskebohunice.sk
[3] http://en.e-obce.sk/obec/jaslovskebohunice/jaslovske-bohunice.html

External links

- Municipal website (http://jaslovskebohunice.sk) (Slovak)

Pressurized_water_reactor

Nuclear Regulatory Commission image of pressurized water reactor vessel heads.

Pressurized water reactors (PWRs) constitute the large majority of all Western nuclear power plants and are one of three types of light water reactor (LWR), the other types being boiling water reactors (BWRs) and supercritical water reactors (SCWRs). In a PWR, the primary coolant (water) is pumped under high pressure to the reactor core where it is heated by the energy generated by the fission of atoms. The heated water then flows to a steam generator where it transfers its thermal energy to a secondary system where steam is generated and flows to turbines which, in turn, spin an electric generator. In contrast to a boiling water reactor, pressure in the primary coolant loop prevents the water from boiling within the reactor. All LWRs use ordinary water as both coolant and neutron moderator.

PWRs were originally designed to serve as nuclear propulsion for nuclear submarines and were used in the original design of the second commercial power plant at Shippingport Atomic Power Station.

PWRs currently operating in the United States are considered Generation II reactors. Russia's VVER reactors are similar to U.S. PWRs. France operates many PWRs to generate the bulk of its electricity.

History

The US Army Nuclear Power Program operated pressurized water reactors from 1954 to 1974.

Three Mile Island Nuclear Generating Station initially operated two pressurized water reactor plants, TMI-1 and TMI-2.[1] The partial meltdown of TMI-2 in 1979 essentially ended the growth in new construction of nuclear power plants in the United States for two decades.[2]

The pressurized water reactor has three new Generation III reactor evolutionary designs: the AP-1000, VVER-1200, ACPR1000+

Design

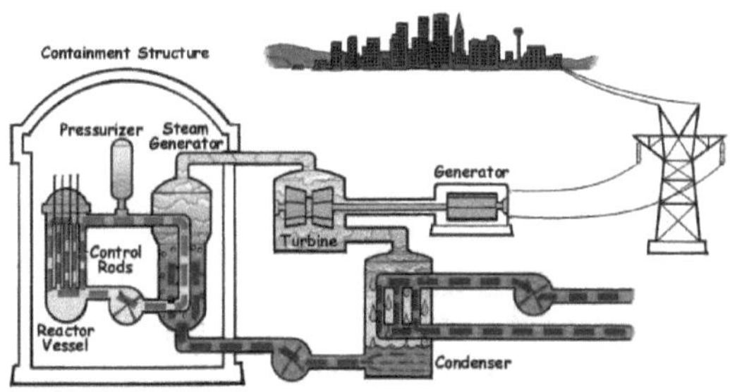

Pictorial explanation of power transfer in a pressurized water reactor. Primary coolant is in orange and the secondary coolant (steam and later feedwater) is in blue.

Nuclear fuel in the reactor vessel is engaged in a fission chain reaction, which produces heat, heating the water in the primary coolant loop by thermal conduction through the fuel cladding. The hot primary coolant is pumped into a heat exchanger called the steam generator, where it flows through hundreds or thousands of tubes (usually 1.9 cm in diameter). Heat is transferred through the walls of these tubes to the lower pressure secondary coolant located on the sheet side of the exchanger where the coolant evaporates to pressurized steam. The transfer of heat is accomplished without mixing the two fluids, which is desirable since the primary coolant might become radioactive. Some common steam generator arrangements are u-tubes or single pass heat exchangers.

In a nuclear power station, the pressurized steam is fed through a steam turbine which drives an electrical generator connected to the electric grid for distribution. After passing through the turbine the secondary coolant (water-steam mixture) is cooled down and condensed in a condenser. The condenser converts the steam to a liquid so that it can be pumped back into the steam generator, and maintains a vacuum at the turbine outlet so that the pressure drop across the turbine, and hence the energy extracted from the steam, is maximized. Before being fed into the steam generator, the condensed steam (referred to as feedwater) is sometimes preheated in order to minimize thermal shock.[3]

The steam generated has other uses besides power generation. In nuclear ships and submarines, the steam is fed through a steam turbine connected to a set of speed reduction gears to a shaft used for propulsion. Direct mechanical action by expansion of the steam can be used for a steam-powered aircraft catapult or similar applications. District heating by the steam is used in some countries and direct heating is applied to internal plant applications.

Two things are characteristic for the pressurized water reactor (PWR) when compared with other reactor types: coolant loop separation from the steam system and pressure inside the primary coolant loop. In a PWR, there are two separate coolant loops (primary and secondary), which are both filled with demineralized/deionized water. A boiling water reactor, by contrast, has only one coolant loop, while more exotic designs such as breeder reactors use substances other than water for coolant and moderator (e.g. sodium in its liquid state as coolant or graphite as a moderator). The pressure in the primary coolant loop is typically 15–16 megapascals (**unknown operator: u'['unknown operator: u'['unknown operator: u'[' unknown operator: u'['**), which is notably higher than in other nuclear reactors, and nearly twice that of a boiling water reactor (BWR). As an effect of this, only localized boiling occurs and steam will recondense promptly in the bulk fluid. By contrast, in a boiling water reactor the primary coolant is designed to boil.[4]

PWR reactor design

Coolant

Light water is used as the primary coolant in a PWR. It enters the bottom of the reactor core at about 275 °C (530 °F) and is heated as it flows upwards through the reactor core to a temperature of about 315 °C (600 °F). The water remains liquid despite the high temperature due to the high pressure in the primary coolant loop, usually around 155 bar (15.5 MPa 153 atm, 2,250 psig). In water, the critical point occurs at around 647 K (374 °C or 705 °F) and 22.064 MPa (3200 PSIA or 218 atm).[5]

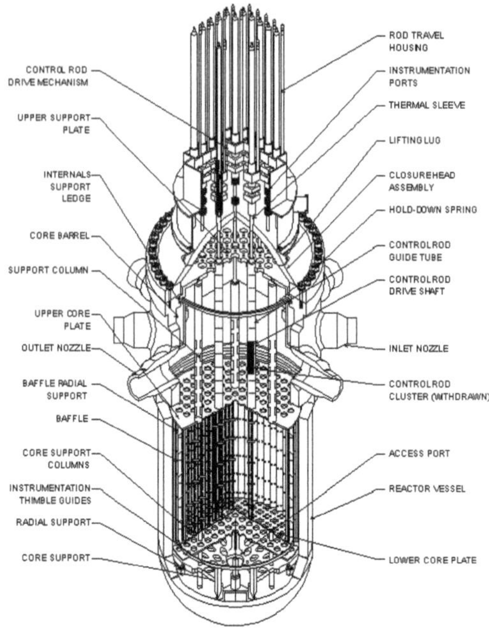

PWR reactor vessel

Pressurizer

Pressure in the primary circuit is maintained by a pressurizer, a separate vessel that is connected to the primary circuit and partially filled with water which is heated to the saturation temperature (boiling point) for the desired pressure by submerged electrical heaters. To achieve a pressure of 155 bar, the pressurizer temperature is maintained at 345 °C (653 °F), which gives a subcooling margin (the difference between the pressurizer temperature and the highest temperature in the reactor core) of 30 °C (54 °F). Thermal transients in the reactor coolant system result in large swings in pressurizer liquid volume, total pressurizer volume is designed around absorbing these transients without uncovering the heaters or emptying the pressurizer. Pressure transients in the primary coolant system manifest as temperature transients in the pressurizer and are controlled through the use of automatic heaters and water spray, which raise and lower pressurizer temperature, respectively.[6]

Pumps

The coolant is pumped around the primary circuit by powerful pumps, which can consume up to 6 MW each.[7] After picking up heat as it passes through the reactor core, the primary coolant transfers heat in a steam generator to water in a lower pressure secondary circuit, evaporating the secondary coolant to saturated steam — in most designs 6.2 MPa (60 atm, 900 psia), 275 °C (530 °F) — for use in the steam turbine. The cooled primary coolant is then returned to the reactor vessel to be heated again.

Moderator

Pressurized water reactors, like all thermal reactor designs, require the fast fission neutrons to be slowed down (a process called moderation or thermalization) in order to interact with the nuclear fuel and sustain the chain reaction. In PWRs the coolant water is used as a moderator by letting the neutrons undergo multiple collisions with light hydrogen atoms in the water, losing speed in the process. This "moderating" of neutrons will happen more often when the water is denser (more collisions will occur). The use of water as a moderator is an important safety feature of PWRs, as an increase in temperature may cause the water to expand, giving greater 'gaps' between the water molecules and reducing the probability of thermalisation - thereby reducing the extent to which neutrons are slowed down and hence reducing the reactivity in the reactor. Therefore, if reactivity increases beyond normal, the reduced moderation of neutrons will cause the chain reaction to slow down, producing less heat. This property, known as the negative temperature coefficient of reactivity, makes PWR reactors very stable. This process is referred to as 'Self-Regulating', i.e. the hotter the coolant becomes, the less reactive the plant becomes, shutting itself down slightly to compensate and vice versa. Thus the plant controls itself around a given temperature set by the position of the control rods.

In contrast, the RBMK reactor design used at Chernobyl, which uses graphite instead of water as the moderator and uses boiling water as the coolant, has a large positive thermal coefficient of reactivity, that increases heat generation when coolant water temperatures increase. This makes the RBMK design less stable than pressurized water reactors. In addition to its property of slowing down neutrons when serving as a moderator, water also has a property of absorbing neutrons, albeit to a lesser degree. When the coolant water temperature increases, the boiling increases, which creates voids. Thus there is less water to absorb thermal neutrons that have already been slowed down by the graphite moderator, causing an increase in reactivity. This property is called the void coefficient of reactivity, and in an RBMK reactor like Chernobyl, the void coefficient is positive, and fairly large, causing rapid transients. This design characteristic of the RBMK reactor is generally seen as one of several causes of the Chernobyl disaster.[8]

Heavy water has very low neutron absorption, so heavy water reactors such as CANDU reactors also have a positive void coefficient, though it is not as large as that of an RBMK like Chernobyl; these reactors are designed with a number of safety systems not found in the original RBMK design, which are designed to handle or react to this as needed.

PWRs are designed to be maintained in an undermoderated state, meaning that there is room for increased water volume or density to further increase moderation, because if moderation were near saturation, then a reduction in density of the moderator/coolant could reduce neutron absorption significantly while reducing moderation only slightly, making the void coefficient positive. Also, light water is actually a somewhat stronger moderator of neutrons than heavy water, though heavy water's neutron absorption is much lower. Because of these two facts, light water reactors have a relatively small moderator volume and therefore have compact cores. One next generation design, the supercritical water reactor, is even less moderated. A less moderated neutron energy spectrum does worsen the capture/fission ratio for ^{235}U and especially ^{239}Pu, meaning that more fissile nuclei fail to fission on neutron absorption and instead capture the neutron to become a heavier nonfissile isotope, wasting one or more neutrons and increasing accumulation of heavy transuranic actinides, some of which have long half-lives.

Fuel

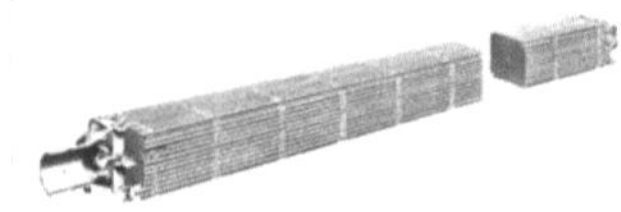

PWR fuel bundle This fuel bundle is from a pressurized water reactor of the nuclear passenger and cargo ship NS *Savannah*. Designed and built by the Babcock and Wilcox Company.

After enrichment, the uranium dioxide (UO_2) powder is fired in a high-temperature, sintering furnace to create hard, ceramic pellets of enriched uranium dioxide. The cylindrical pellets are then clad in a corrosion-resistant zirconium metal alloy Zircaloy which are backfilled with helium to aid heat conduction and detect leakages. Zircaloy is chosen because of its mechanical properties and its low absorption cross section.[9] The finished fuel rods are grouped in fuel assemblies, called fuel bundles, that are then used to build the core of the reactor. A typical PWR has fuel assemblies of 200 to 300 rods each, and a large reactor would have about 150–250 such assemblies with 80–100 tonnes of uranium in all. Generally, the fuel bundles consist of fuel rods bundled 14 × 14 to 17 × 17. A PWR produces on the order of 900 to 1,600 MW_e. PWR fuel bundles are about 4 meters in length.[10]

Refuelings for most commercial PWRs is on an 18–24 month cycle. Approximately one third of the core is replaced each refueling, though some more modern refueling schemes may reduce refuel time to a few days and allow refueling to occur on a shorter periodicity.[11]

Control

In PWRs reactor power can be viewed as following steam (turbine) demand due to the reactivity feedback of the temperature change caused by increased or decreased steam flow. (See: Negative temperature coefficient.) Boron and control rods are used to maintain primary system temperature at the desired point. In order to decrease power, the operator throttles shut turbine inlet valves. This would result in less steam being drawn from the steam generators. This results in the primary loop increasing in temperature. The higher temperature causes less fission and decreases power. This decrease of power will eventually result in primary system temperature returning to its previous steady state value. The operator can control the steady state operating temperature by addition of boric acid and/or movement of control rods.

Reactivity adjustment to maintain 100% power as the fuel is burned up in most commercial PWRs is normally achieved by varying the concentration of boric acid dissolved in the primary reactor coolant. Boron readily absorbs neutrons and increasing or decreasing its concentration in the reactor coolant will therefore affect the neutron activity correspondingly. An entire control system involving high pressure pumps (usually called the charging and letdown system) is required to remove water from the high pressure primary loop and re-inject the water back in with differing concentrations of boric acid. The reactor control rods, inserted through the reactor vessel head directly into the fuel bundles, are moved for the following reasons:

- To start up the reactor.
- To shut down the primary nuclear reactions in the reactor.
- To accommodate short term transients such as changes to load on the turbine.

The control rods can also be used:

- To compensate for nuclear poison inventory.
- To compensate for nuclear fuel depletion.

but these effects are more usually accommodated by altering the primary coolant boric acid concentration.

In contrast, BWRs have no boron in the reactor coolant and control the reactor power by adjusting the reactor coolant flow rate

Advantages

- PWR reactors are very stable due to their tendency to produce less power as temperatures increase; this makes the reactor easier to operate from a stability standpoint.
- PWR turbine cycle loop is separate from the primary loop, so the water in the secondary loop is not contaminated by radioactive materials.
- PWRs can passively scram the reactor in the event that offsite power is lost to immediately stop the primary nuclear reaction. The control rods are held by electromagnets and fall by gravity when current is lost; full insertion safely shuts down the primary nuclear reaction.
- PWR technology is favoured by nations seeking to develop a nuclear navy, the compact reactors fit well in nuclear submarines and other nuclear ships.

Disadvantages

- The coolant water must be highly pressurized to remain liquid at high temperatures. This requires high strength piping and a heavy pressure vessel and hence increases construction costs. The higher pressure can increase the consequences of a loss-of-coolant accident.[12] The reactor pressure vessel is manufactured from ductile steel but, as the plant is operated, neutron flux from the reactor causes this steel to become less ductile. Eventually the ductility of the steel will reach limits determined by the applicable boiler and pressure vessel standards, and the pressure vessel must be repaired or replaced. This might not be practical or economic, and so determines the life of the plant.
- Additional high pressure components such as reactor coolant pumps, pressurizer, steam generators, etc. are also needed. This also increases the capital cost and complexity of a PWR power plant.
- The high temperature water coolant with boric acid dissolved in it is corrosive to carbon steel (but not stainless steel); this can cause radioactive corrosion products to circulate in the primary coolant loop. This not only limits the lifetime of the reactor, but the systems that filter out the corrosion products and adjust the boric acid concentration add significantly to the overall cost of the reactor and to radiation exposure. In one instance, this has resulted in severe corrosion to control rod drive mechanisms when the boric acid solution leaked through the seal between the mechanism itself and the primary system.[13] [14]
- Natural uranium is only 0.7% uranium-235, the isotope necessary for thermal reactors. This makes it necessary to enrich the uranium fuel, which increases the costs of fuel production. If heavy water is used, it is possible to operate the reactor with natural uranium, but the production of heavy water requires large amounts of energy and is hence expensive.
- Because water acts as a neutron moderator, it is not possible to build a fast neutron reactor with a PWR design. A reduced moderation water reactor may however achieve a breeding ratio greater than unity, though this reactor design has disadvantages of its own.[15]

See also

- Boiling water reactor
- List of PWR reactors
- Nuclear safety systems

Next generation designs

- Areva EPR
- Westinghouse Advanced Passive 1000 (AP1000)
- Rosatom VVER-1200 (or NPP-2006)
- China Guangdong Nuclear Power Company ACPR-1000 (or ACPR1000+)

Notes

[1] Mosey 1990, pp.69–71
[3] Glasstone & Senonske 1994, pp. 769
[4] Duderstadt & Hamilton 1976, pp. 91–92
[5] International Association for the Properties of Water and Steam, 2007.
[6] Glasstone & Senonske 1994, pp. 767
[7] Tong 1988, pp. 175
[8] Mosey 1990, pp. 92–94
[10] Glasstone & Sesonske 1994, pp. 21
[11] Duderstadt & Hamilton 1976, pp. 598
[12] Tong 1988, pp. 216–217
[15] Duderstadt & Hamilton 1976, pp. 86

External links

- Nuclear Science and Engineering (http://ocw.mit.edu/courses/nuclear-engineering/) at MIT OpenCourseWare.
- Document archives (http://www.nrc.gov/reading-rm/doc-collections/) at the website of the United States Nuclear Regulatory Commission.
- Operating Principles of a Pressurized Water Reactor (http://www.youtube.com/watch?v=MSFgmLW1Crw) (YouTube video).

VVER

The **VVER**, or **WWER**, (from Russian: Водо-водяной энергетический реактор; transliterates as *Vodo-Vodyanoi Energetichesky Reactor*; *Water-Water Power Reactor*) is a series of pressurised water reactor designs originally developed in the Soviet Union, and now Russia, by OKB Gidropress.[] Power output ranges from 300 MWe to 1700 MWe[] with the latest Russian development of the design. VVER power stations are used by Armenia, Bulgaria, China, Czech Republic, Finland, Hungary, India, Iran, Slovakia, Ukraine, and the Russian Federation.

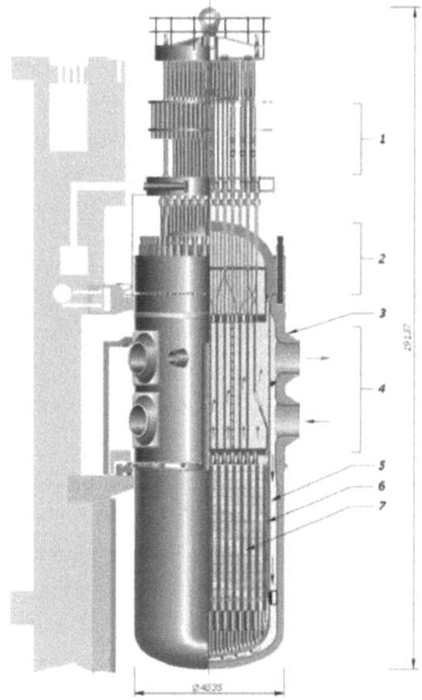

WWER-1000 (or VVER-1000 as a direct transliteration of Russian ВВЭР-1000) is a 1000 MWe Russian nuclear power reactor of PWR type.

History

The earliest VVERs were built before 1970. The VVER-440 Model V230 was the most common design, delivering 440 MW of electrical power. The V230 employs six primary coolant loops each with a horizontal steam generator. A modified version of VVER-440, Model V213, was a product of the first nuclear safety standards adopted by Soviet designers. This model includes added emergency core cooling and auxiliary feedwater systems as well as upgraded accident localization systems.[]

The larger VVER-1000 was developed after 1975 and is a four-loop system housed in a containment-type structure with a spray steam suppression system. VVER reactor designs have been elaborated to incorporate automatic control, passive safety and containment systems associated with Western third generation nuclear reactors.

The VVER-1200 is the version currently offered for construction, being an evolution of the VVER-1000 with increased power output to about 1200 MWe (gross) and providing additional passive safety features.[]

In 2012 Rosatom stated that in the future it intended to certify the VVER with the British and U.S. regulatory authorities.[1]

Design

The Russian abbreviation VVER stands for 'water-water energy reactor' (i.e. water-cooled water-moderated energy reactor). This describes the pressurised water reactor (PWR) design. The main distinguishing features of the VVER[] compared to other PWRs are:

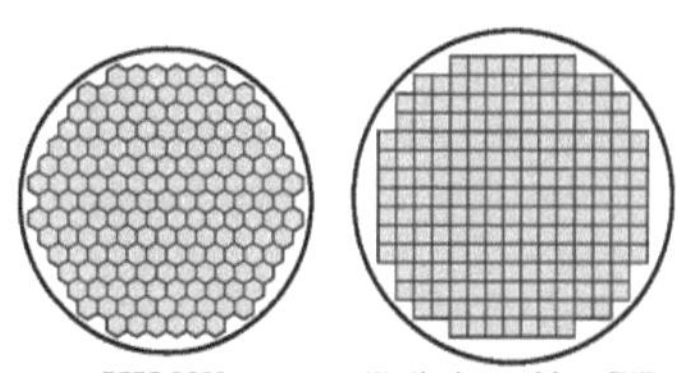

The arrangement of hexahedral fuel assemblies compared to a Westinghouse PWR design

- Horizontal steam generators
- Hexagonal fuel assemblies
- No bottom penetrations in the pressure vessel
- High-capacity pressurisers providing a large reactor coolant inventory

Reactor fuel rods are fully immersed in water kept at 15 MPa of pressure so that it does not boil at normal (220 to over 300 °C) operating temperatures. Water in the reactor serves both as a coolant and a moderator which is an important safety feature. Should coolant circulation fail the neutron moderation effect of the water diminishes, reducing reaction intensity and compensating for loss of cooling, a condition known as negative void coefficient. Later versions of the reactors are encased in massive steel pressure shells. Fuel is low enriched (ca. 2.4–4.4% ^{235}U) uranium dioxide (UO_2) or equivalent pressed into pellets and assembled into fuel rods.

Reactivity is controlled by control rods that can be inserted into the reactor from above. These rods are made from a neutron absorbing material and depending on depth of insertion hinder the chain reaction. If there is an emergency, a reactor shutdown can be performed by full insertion of the control rods into the core.

Primary cooling circuit

As stated above, water in the primary circuit is kept under constant pressure to avoid boiling. Since the water transfers all the heat from the core and is irradiated, integrity of this circuit is most crucial. In the circuit four subsystems can be distinguished:

1. Reactor: Water flows through fuel rod assemblies and is heated by the nuclear chain reaction.
2. Volume compensator: To keep the water under constant but controlled pressure, the volume compensator regulates pressure employing self-regulation of saturated steam-water interface and by means of electrical heating and relief valves.
3. Steam Generator: In the steam generator, heat from primary coolant water is used to boil water in the secondary circuit.
4. Pump: The pump ensures proper circulation of the water through the circuit.

To ensure safety primary components are redundant.

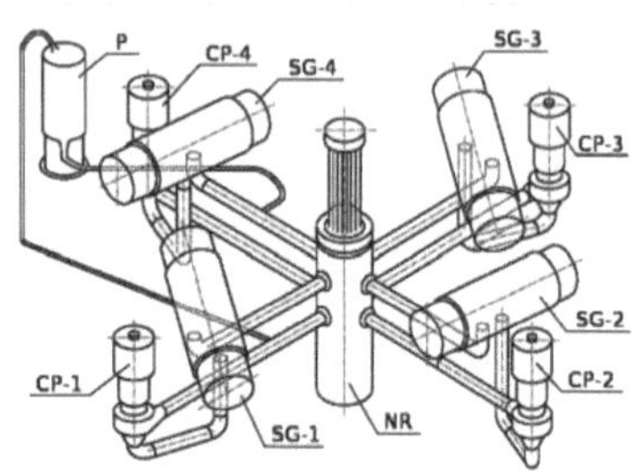

Stereometric scheme of primary cooling circuit VVER-1000

Treatment of the interior part of a VVER-1000 reactor frame on Atommash

Secondary circuit and electrical output

The secondary circuit also consists of different subsystems:

1. Steam Generator: Secondary water is boiled taking heat from the primary circuit. Before entering the turbine remaining water is separated from the steam so that the steam is dry.
2. Turbine: The expanding steam drives a turbine, which connects to an electrical generator. The turbine is split into high and low pressure sections. To prevent condensation (Water droplets at high speed damage the turbine blades) steam is reheated between these sections. Reactors of the VVER-1000 type deliver 1 GW of electrical power.
3. Condenser: The steam is cooled and allowed to condense, shedding waste heat into a cooling circuit.
4. Deaerator: Removes gases from the coolant.
5. Pump: The circulation pumps are each driven by their own small steam turbine.

To increase efficiency of the process, steam from the turbine is taken to reheat coolant before the deaerator and the steam generator. Water in this circuit is not supposed to be radioactive.

Cooling circuit

The cooling circuit is an open circuit diverting water from an outside reservoir such as a lake or river. Evaporative cooling towers, cooling basins or ponds exhaust waste heat from the generation circuit, releasing it into the environment. In addition to generating electricity most VVERs have a capability to supply heat for residential and industrial use. Operational examples of such systems are the plants at Jaslovské Bohunice and Dukovany. [2]

Safety barriers

A typical design feature of nuclear reactors is layered safety barriers preventing escape of radioactive material. VVER reactors have four layers:

1. Fuel pellets: Radioactive elements are retained within the crystal structure of the fuel pellets.
2. Fuel rods: The zircaloy tubes provide a further barrier resistant to heat and high pressure.
3. Reactor Shell: A massive steel shell encases the whole fuel assembly hermetically.
4. Reactor Building: A concrete containment building that encases the whole first circuit is strong enough to resist the pressure surge a breach in the first circuit would cause.

The two VVER-440 units in Loviisa have containment buildings that fulfil Western safety standards.

Currently operating Russian VVERs are inherently safer designs than the RBMK reactors of Chernobyl disaster. The Soviet Union opted to construct graphite-moderated RBMK series nuclear reactors without containment structures on grounds of cost as well as the relative ease of re-fueling RBMK reactors. Fuel elements in a RBMK reactor can be replaced while still operational, allowing continued operation and plutonium extraction compared to the VVER which needs to be shut down. Many levels of protection and containment have both been proposed and constructed for RBMK and VVER type reactors.

Operational life of VVER 1000

When first built the VVER design was intended to be operational for 35 years. A mid-life major overhaul including a complete replacement of critical parts such as fuel and control rod channels was thought necessary after that.[] Since RBMK reactors specified a major replacement programme at 35 years designers originally decided this needed to happen in the VVER type as well, although they are of more robust design than the RBMK type. Most of Russia's VVER plants are now reaching and passing the 35 year mark. More recent design studies have allowed for an extension of lifetime up to 50 years with replacement of equipment. New VVERs will be nameplated with the extended lifetime.

Control room of a VVER-1000 in 2009, Kozloduy Unit 5

In 2010 the oldest VVER-1000, at Novovoronezh, was shut down for modernization to extend its operating life for an additional 30 years; the first to undergo such an operating life extension. The works include the modernization of management, protection and emergency systems, and improvement of security and radiation safety systems.[3]

VVER-1200

The VVER-1200 (or NPP-2006 or AES-2006)[] is an evolution of the VVER-1000 being offered for domestic and export use.[] [] Specifications include a $1,200 per kW electric capital cost, 54 month planned construction time, and expected 50 year lifetime at 90% capacity factor. The VVER 1200 will produce 1,200 MWe of power. Safety features include a containment building and missile shield. It will have full emergency systems that include an emergency core cooling system, emergency backup diesel power supply, advanced refueling machine, computerized reactor control systems, backup feedwater supply and reactor SCRAM system. The nuclear reactor and associated systems will be hosted in one single building and there will be another building for the turbogenerators. The main building will comprise the reactor, refueling machine and diesel backup power supply, steam generators and reactor control systems.

If a VVER-1200 experiences a loss of coolant accident or loss of power accident the turbogenerators 'coast down' for 30 seconds, during which time a shutdown can be initiated using residual power in the system. Further emergency power is available from a backup set of diesel generators kept on standby to maintain cooling flow to the reactor. The reactor design has been refined to optimize fuel efficiency.

The first two units are proposed for Leningrad Nuclear Power Plant II and Novovoronezh Nuclear Power Plant II. A standardized design has not been elected. Many more reactors with a VVER-1200/491 like the Leningrad-II-design are firmly planned (Kaliningrad and Nizhny Novgorod NPP) and under construction. The VVER-1200/392M[] under construction at the Novovoronezh NPP-II is selected for the Seversk, Zentral and South-Urals NPP. A standard version was developed as VVER-1200/510 and referred to as VVER-TOI.

In July 2012 a contract was agreed to build two AES-2006 in Belarus at Ostrovets for a cost of about $10 billion.[]

Passive heat removal system

A passive heat removal system has been added to the existing active systems in the AES-92 version of the VVER-1000 used for the Koodankulam Nuclear Power Plant in India. This has been retained for the newer VVER-1200 and future designs. The system is based on a cooling system and water tanks built on top of the containment dome.[4] The passive systems all safety functions for 24 hours, and core safety for 72 hours.[]

Other new safety systems include aircraft crash protection and system to contain the molten reactor core in the event of a severe accident.[] []

Future versions

A number of designs for future versions of the VVER have been made:[]

- MIR-1200 (Modernised International Reactor) - designed in conjunction with Czech company ŠKODA JS[] to satisfy European requirements[]
- VVER-1500 - VVER-1000 with dimensions increased to produce 1500 MWe gross power output, but design shelved in favour of the evolutionary VVER-1200[]
- VVER-TOI is aimed at development of typical optimized informative-advanced project of a new generation III+ Power Unit based on VVER technology, which meets a number of target-oriented parameters using modern information and management technologies.[5]

Power plants

List of operational VVER installations

Power plant	Country	Reactors	Notes
Akkuyu	Turkey	(4 × VVER-1200/491) (AES-2006)	Under construction.
Balakovo	Russia	4 × VVER-1000/320 (2 × VVER-1000/320)	Unit 5 and 6 construction suspended.
Belene	Bulgaria	(2 × VVER-1000/446)	Planned.[]
Bohunice	Slovakia	2 × VVER-440/230 2 × VVER-440/213	Split in two plants, V-1 and V-2 with two reactors each. VVER-440/230 units decommissioned in 2007.
Bushehr	Iran	1 × VVER-1000/446 (3 × VVER-1000/446)	A version of the V-320 adapted to the Bushehr site.[] Unit 2 and 3 planned, unit 4 cancelled.
Dukovany	Czech Republic	4 × VVER 440/213	Now upgraded to 502 MW in 2009-2012.
Greifswald	Germany	4 × VVER-440/230 1 × VVER-440/213 (3 × VVER-440/213)	Decommissioned. Unit 6 finished, but never operated. Unit 7 and 8 construction suspended.
Kalinin	Russia	2 × VVER-1000/338 1 × VVER-1000/320 (1 × VVER-1000/320)	Unit 4 under construction, operational 2011.
Khmelnitskiy	Ukraine	2 × VVER-1000/320 (2 × VVER-1000/392B)	Unit 3 and 4 under construction.
Kola	Russia	2 × VVER-440/230 2 × VVER-440/213	
Koodankulam	India	(2 × VVER-1000/412) (AES-92)	Under construction, operational 2012/2013 with four additional units planned.
Kozloduy	Bulgaria	4 × VVER-440/230 2 × VVER-1000	VVER-440/230 units decommissioned 2004-2007.
Leningrad II	Russia	2 × VVER-1200/491 (2 × VVER-1200/491)	The units are the prototypes of the VVER-1200/491 (AES-2006) and under construction.
Loviisa	Finland	2 × VVER-440/213	Western control systems, Totally different containment structures. Later modified for a 496 MW output.
Metsamor	Armenia	2 × VVER-440/230	One reactor was shut down in 1989.
Mochovce	Slovakia	2 × VVER-440/213 (2 × VVER-440/213)	Units 3 and 4 construction suspended due to lack of funds, planned to be operational in 2012.
Novovoronezh	Russia	1 x VVER-210 (V-1) 1 x VVER-365 (V-3) 2 × VVER-440/179 1 × VVER-1000/187	All units are prototypes. Unit 1 and 2 shutdown. Unit 3 modernised in 2002.[6]
Novovoronezh II	Russia	(2 × VVER-1200/392M) (AES-2006)	The units are the prototypes of the VVER-1200/392M (AES-2006) and under construction.
Paks	Hungary	4 × VVER-440/213	Two VVER-1000/320 plan was cancelled.
Rheinsberg	Germany	1 × VVER-210	Unit decommissioned

Rivne	Ukraine	2 × VVER-440/213 2 × VVER-1000/320 (2 × VVER-1000/320)	Unit 5 and 6 planning suspended.
South Ukraine	Ukraine	1 × VVER-1000/302 1 × VVER-1000/338 1 × VVER-1000/320 (1 × VVER-1000/320)	unit 4 construction suspended.
Stendal	Germany	(4 × VVER-1000/320)	All 4 units construction cancelled after Germany reunification.
Temelin	Czech Republic	2 × VVER-1000/320 (2 × VVER-1000/320)	Unit 3 and 4 construction suspended. Now unit 3 and 4 in planning again (operated in 2025).
Tianwan	China	2 × VVER-1000/428 (AES-91) (6 × VVER-1000/428)	Unit 3 to 8 firmly planned.
Volgodonsk	Russia	2 × VVER-1000/320 (2 × VVER-1000/320)	Unit 3 and 4 is under construction and planned to be operational in 2013 and 2014.
Zaporizhzhia	Ukraine	6 × VVER-1000/320	Largest nuclear power plant in Europe.

See the Wikipedia pages for each facility for sources.

Russia recently installed two nuclear reactors in China at the Tianwan Nuclear Power Plant, and an extension consisting of a further two reactors was just approved. This is the first time the two countries have co-operated on a nuclear power project. The reactors are the VVER 1000 type, which Russia has improved incrementally while retaining the basic design. These VVER 1000 reactors are housed in a confinement shell capable of being hit by an aircraft weighing 20 tonnes and suffering no expected damage. Other important safety features include an emergency core cooling system and core confinement system. Russia delivered initial fuel loads for the Tianwan reactors. China planned to begin indigenous fuel fabrication for the Tianwan plant in 2010, using technology transferred from Russian nuclear fuel producer TVEL.[7]

The Tianwan Nuclear Power Plant uses many third party parts. While the reactor and turbo-generators are of Russian design, the control room was designed and built by an international consortium. In this way the plant was brought to meet widely recognised safety standards; safety systems were already mostly in place but the previous monitoring of these systems did not meet international safety standards. The new VVER 1000 plant built in China has 94% of its systems automated, meaning the plant can control itself under most situations. Refueling procedures require little human intervention. Five operators are still needed in the control room. The IAEA has referred to the station as the "safest nuclear power plant in the world".[8]

In May 2010 Russia secured an agreement with the Turkish government to build a power plant with four VVER-1200 reactors at Akkuyu, Turkey.[9] [10] However, due to the accident experienced in Fukushima, anti-nuclear environmentalist groups heavily protested the proposed reactor at Akkuyu.

On 11 October 2011 an agreement was signed to build Belarus' first nuclear power plant at Ostrovets, using two NPP-2006 reactors with active and passive safety systems. The first unit is planned to be completed by 2017.[11]

Gallery

NPP Balakovo VVER-1000

NPP Rovno VVER-1000

NPP Chmelnyckyj VVER-1000

NPP Balakovo VVER-1000

NPP Leningrad VVER-1200

NPP Novovoroněž VVER-1200

NPP Novovoroněž VVER-1200

NPP Kudankulam VVER-1000

NPP Stendal VVER-1000

NPP Żarnowiec VVER-440

NPP Juragua VVER-440

NPP Greifswald VVER-440

See also

- Russian floating nuclear power station
- VBER-300

References

[2] Cogeneration in the Former Soviet Union; June 24 1997 (http://www.osti.gov/bridge/servlets/purl/493376-jkHJc9/webviewable/493376.pdf;)

[7] Tianwan fuel fabrication moves to China; March 2010 (http://www.world-nuclear-news.org/ENF-Tianwan_fuel_fabrication_moves_to_China-0203104.html)

[8] Russian-Chinese nuclear station safest in the world: IAEA (http://rt.com/news/russian-chinese-nuclear-station-safest-in-the-world-iaea/), RussiaToday, 2007-12-07

External links

- WWER-type reactor plants (http://www.gidropress.podolsk.ru/en/projects/wwer.php), OKB Gidropress
- AES-2006 (VVER-1200) (http://www.old.rosatom.ru/en/about/projects/npp_2006/), Rosatom
- VVER Reactor (http://www.nucleartourist.com/type/vver.htm), Virtual Nuclear Tourist

Electricity

Lightning is one of the most dramatic effects of electricity.

Electricity is the set of physical phenomena associated with the presence and flow of electric charge. Electricity gives a wide variety of well-known effects, such as lightning, static electricity, electromagnetic induction and the flow of electrical current. In addition, electricity permits the creation and reception of electromagnetic radiation such as radio waves.

In electricity, charges produce electromagnetic fields which act on other charges. Electricity occurs due to several types of physics:

- **electric charge**: a property of some subatomic particles, which determines their electromagnetic interactions. Electrically charged matter is influenced by, and produces, electromagnetic fields.
- **electric current**: a movement or flow of electrically charged particles, typically measured in amperes.
- **electric field** (see electrostatics): an especially simple type of electromagnetic field produced by an electric charge even when it is not moving (i.e., there is no electric current). The electric field produces a force on other charges in its vicinity. Moving charges additionally produce a magnetic field.
- **electric potential**: the capacity of an electric field to do work on an electric charge, typically measured in volts.
- **electromagnets**: electrical currents generate magnetic fields, and changing magnetic fields generate electrical currents

In electrical engineering, electricity is used for:

- **electric power** where electric current is used to energise equipment
- **electronics** which deals with electrical circuits that involve active electrical components such as vacuum tubes, transistors, diodes and integrated circuits, and associated passive interconnection technologies.

Electrical phenomena have been studied since antiquity, though advances in the science were not made until the seventeenth and eighteenth centuries. Practical applications for electricity however remained few, and it would not be until the late nineteenth century that engineers were able to put it to industrial and residential use. The rapid expansion in electrical technology at this time transformed industry and society. Electricity's extraordinary versatility as a means of providing energy means it can be put to an almost limitless set of applications which include transport, heating, lighting, communications, and computation. Electrical power is the backbone of modern industrial society.[1]

The word *electricity* is from the New Latin *ēlectricus*, "amber-like"[a], coined in the year 1600 from the Greek *ήλεκτρον* (electron) meaning amber, because electrical effects were produced classically by rubbing amber.

History

Main articles: History of electromagnetic theory and History of electrical engineering. See also: Etymology of electricity

Thales, the earliest known researcher into electricity

Long before any knowledge of electricity existed people were aware of shocks from electric fish. Ancient Egyptian texts dating from 2750 BC referred to these fish as the "Thunderer of the Nile", and described them as the "protectors" of all other fish. Electric fish were again reported millennia later by ancient Greek, Roman and Arabic naturalists and physicians.[2] Several ancient writers, such as Pliny the Elder and Scribonius Largus, attested to the numbing effect of electric shocks delivered by catfish and torpedo rays, and knew that such shocks could travel along conducting objects.[] Patients suffering from ailments such as gout or headache were directed to touch electric fish in the hope that the powerful jolt might cure them.[3] Possibly the earliest and nearest approach to the discovery of the identity of lightning, and electricity from any other source, is to be attributed to the Arabs, who before the 15th century had the Arabic word for lightning (*raad*) applied to the electric ray.[4]

Ancient cultures around the Mediterranean knew that certain objects, such as rods of amber, could be rubbed with cat's fur to attract light objects like feathers. Thales of Miletos made a series of observations on static electricity around 600 BC, from which he believed that friction rendered amber magnetic, in contrast to minerals such as magnetite, which needed no rubbing.[] [5] Thales was incorrect in believing the attraction was due to a magnetic effect, but later science would prove a link between magnetism and electricity. According to a controversial theory, the Parthians may have had knowledge of electroplating, based on the 1936 discovery of the Baghdad Battery, which resembles a galvanic cell, though it is uncertain whether the artifact was electrical in nature.[6]

Benjamin Franklin conducted extensive research on electricity in the 18th century, as documented by Joseph Priestley (1767) *History and Present Status of Electricity*, with whom Franklin carried on extended correspondence.

Electricity would remain little more than an intellectual curiosity for millennia until 1600, when the English scientist William Gilbert made a careful study of electricity and magnetism, distinguishing the lodestone effect from static electricity produced by rubbing amber.[] He coined the New Latin word *electricus* ("of amber" or "like amber", from ἤλεκτρον [*elektron*], the Greek word for "amber") to refer to the property of attracting small objects after being rubbed.[7] This association gave rise to the English words "electric" and "electricity", which made their first appearance in print in Thomas Browne's *Pseudodoxia Epidemica* of 1646.[8]

Further work was conducted by Otto von Guericke, Robert Boyle, Stephen Gray and C. F. du Fay. In the 18th century, Benjamin Franklin conducted extensive research in electricity, selling his possessions to fund his work. In June 1752 he is reputed to have attached a metal key to the bottom of a dampened kite string and flown the kite in a storm-threatened sky.[9] A succession of sparks jumping from the key to the back of his hand showed that lightning was indeed electrical in nature.[10] He also explained the apparently paradoxical behavior of the Leyden jar as a device for storing large amounts of electrical charge.

Michael Faraday formed the foundation of electric motor technology

In 1791, Luigi Galvani published his discovery of bioelectricity, demonstrating that electricity was the medium by which nerve cells passed signals to the muscles.[] Alessandro Volta's battery, or voltaic pile, of 1800, made from alternating layers of zinc and copper, provided scientists with a more reliable source of electrical energy than the electrostatic machines previously used.[] The recognition of electromagnetism, the unity of electric and magnetic phenomena, is due to Hans Christian Ørsted and André-Marie Ampère in 1819-1820; Michael Faraday invented the electric motor in 1821, and Georg Ohm mathematically analysed the electrical circuit in 1827.[] Electricity and magnetism (and light) were definitively linked by James Clerk Maxwell, in particular in his "On Physical Lines of Force" in 1861 and 1862.[11]

While the early 19th century had seen rapid progress in electrical science, the late 19th century would see the greatest progress in electrical engineering. Through such people as Nikola Tesla, Galileo Ferraris, Oliver Heaviside, Thomas Edison, Ottó Bláthy, Ányos Jedlik, Sir Charles Parsons, Joseph Swan, George Westinghouse, Ernst Werner von Siemens, Alexander Graham Bell and Lord Kelvin, electricity turned from a scientific curiosity into an essential tool for modern life, becoming a driving force of the Second Industrial Revolution.[12]

In 1887, Heinrich Hertz[13] [] discovered that electrodes illuminated with ultraviolet light create electric sparks more easily. In 1905 Albert Einstein published a paper that explained experimental data from the photoelectric effect as being the result of light energy being carried in discrete quantized packets, energising electrons. This discovery led to the quantum revolution. Einstein was awarded the Nobel Prize in 1921 for "his discovery of the law of the photoelectric effect".[14] The photoelectric effect is also employed in photocells such as can be found in solar panels and this is frequently used to make electricity commercially.

The first solid-state device was the "cat's whisker" detector, first used in 1930s radio receivers. A whisker-like wire is placed lightly in contact with a solid crystal (such as a germanium crystal) in order to detect a radio signal by the contact junction effect.[15] In a solid-state component, the current is confined to solid elements and compounds engineered specifically to switch and amplify it. Current flow can be understood in two forms: as negatively charged electrons, and as positively charged electron deficiencies called holes. These charges and holes are understood in terms of quantum physics. The building material is most often a crystalline semiconductor.[16] [17]

The solid-state device came into its own with the invention of the transistor in 1947. Common solid-state devices include transistors, microprocessor chips, and RAM. A specialized type of RAM called flash RAM is used in flash drives and more recently, solid state drives to replace mechanically rotating magnetic disc hard drives. Solid state devices became prevalent in the 1950s and the 1960s, during the transition from vacuum tube technology to semiconductor diodes, transistors, integrated circuit (IC) and the light-emitting diode (LED).

Concepts

Electric charge

Main article: Electric charge. See also: electron, proton, and ion.

Electric charge is a property of certain subatomic particles, which gives rise to and interacts with the electromagnetic force, one of the four fundamental forces of nature. Charge originates in the atom, in which its most familiar carriers are the electron and proton. It is a conserved quantity, that is, the net charge within an isolated system will always remain constant regardless of any changes taking place within that system.[18] Within the system, charge may be transferred between bodies, either by direct contact, or by passing along a conducting material, such as a wire.[] The informal term static electricity refers to the net presence (or 'imbalance') of charge on a body, usually caused when dissimilar materials are rubbed together, transferring charge from one to the other.

The presence of charge gives rise to the electromagnetic force: charges exert a force on each other, an effect that was known, though not understood, in antiquity.[] A lightweight ball suspended from a string can be charged by touching it with a glass rod that has itself been charged by rubbing with a cloth. If a similar ball is charged by the same glass rod, it is found to repel the first: the charge acts to force the two balls apart. Two balls that are charged with a rubbed amber rod also repel each other. However, if one ball is charged by the glass rod, and the other by an amber rod, the two balls are found to attract each other. These phenomena were investigated in the late eighteenth century by Charles-Augustin de Coulomb, who deduced that charge manifests itself in two opposing forms. This discovery led to the well-known axiom: *like-charged objects repel and opposite-charged objects attract.*[]

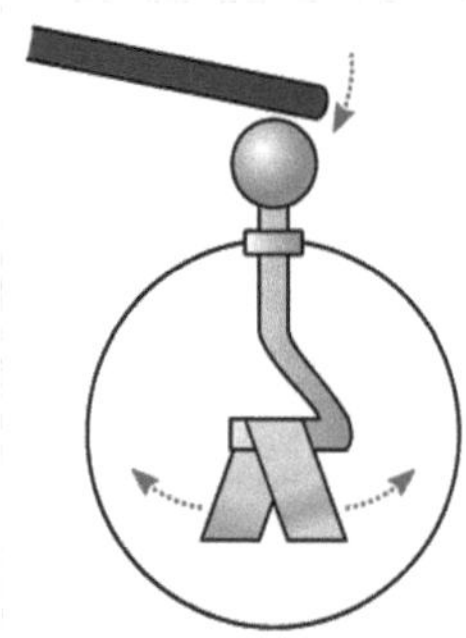

Charge on a gold-leaf electroscope causes the leaves to visibly repel each other

The force acts on the charged particles themselves, hence charge has a tendency to spread itself as evenly as possible over a conducting surface. The magnitude of the electromagnetic force, whether attractive or repulsive, is given by Coulomb's law, which relates the force to the product of the charges and has an inverse-square relation to the distance between them.[19] [20] The electromagnetic force is very strong, second only in strength to the strong interaction,[21] but unlike that force it operates over all distances.[] In comparison with the much weaker gravitational force, the electromagnetic force pushing two electrons apart is 10^{42} times that of the gravitational attraction pulling them together.[]

The charge on electrons and protons is opposite in sign, hence an amount of charge may be expressed as being either negative or positive. By convention, the charge carried by electrons is deemed negative, and that by protons positive, a custom that originated with the work of Benjamin Franklin.[22] The amount of charge is usually given the symbol Q and expressed in coulombs;[23] each electron carries the same charge of approximately -1.6022×10^{-19} coulomb. The proton has a charge that is equal and opposite, and thus $+1.6022\times10^{-19}$ coulomb. Charge is possessed not just by matter, but also by antimatter, each antiparticle bearing an equal and opposite charge to its corresponding particle.[24]

Charge can be measured by a number of means, an early instrument being the gold-leaf electroscope, which although still in use for classroom demonstrations, has been superseded by the electronic electrometer.[]

Electric current

The movement of electric charge is known as an electric current, the intensity of which is usually measured in amperes. Current can consist of any moving charged particles; most commonly these are electrons, but any charge in motion constitutes a current.

By historical convention, a positive current is defined as having the same direction of flow as any positive charge it contains, or to flow from the most positive part of a circuit to the most negative part. Current defined in this manner is called conventional current. The motion of negatively charged electrons around an electric circuit, one of the most familiar forms of current, is thus deemed positive in the *opposite* direction to that of the electrons.[25] However, depending on the conditions, an electric current can consist of a flow of charged particles in either direction, or even in both directions at once. The positive-to-negative convention is widely used to simplify this situation.

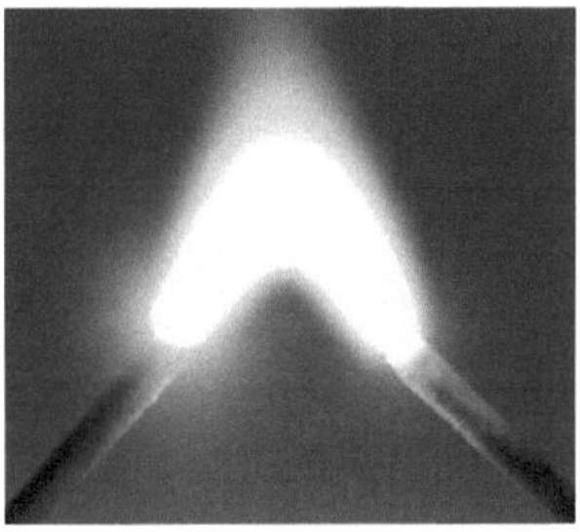

An electric arc provides an energetic demonstration of electric current

The process by which electric current passes through a material is termed electrical conduction, and its nature varies with that of the charged particles and the material through which they are travelling. Examples of electric currents include metallic conduction, where electrons flow through a conductor such as metal, and electrolysis, where ions (charged atoms) flow through liquids. While the particles themselves can move quite slowly, sometimes with an average drift velocity only fractions of a millimetre per second,[] the electric field that drives them itself propagates at close to the speed of light, enabling electrical signals to pass rapidly along wires.[26]

Current causes several observable effects, which historically were the means of recognising its presence. That water could be decomposed by the current from a voltaic pile was discovered by Nicholson and Carlisle in 1800, a process now known as electrolysis. Their work was greatly expanded upon by Michael Faraday in 1833.[] Current through a resistance causes localised heating, an effect James Prescott Joule studied mathematically in 1840.[] One of the most important discoveries relating to current was made accidentally by Hans Christian Ørsted in 1820, when, while preparing a lecture, he witnessed the current in a wire disturbing the needle of a magnetic compass.[27] He had discovered electromagnetism, a fundamental interaction between electricity and magnetics. The level of electromagnetic emissions generated by electric arcing is high enough to produce electromagnetic interference, which can be detrimental to the workings of adjacent equipment.[28]

In engineering or household applications, current is often described as being either direct current (DC) or alternating current (AC). These terms refer to how the current varies in time. Direct current, as produced by example from a battery and required by most electronic devices, is a unidirectional flow from the positive part of a circuit to the negative.[] If, as is most common, this flow is carried by electrons, they will be travelling in the opposite direction. Alternating current is any current that reverses direction repeatedly; almost always this takes the form of a sine wave.[] Alternating current thus pulses back and forth within a conductor without the charge moving any net distance over time. The time-averaged value of an alternating current is zero, but it delivers energy in first one direction, and then the reverse. Alternating current is affected by electrical properties that are not observed under steady state direct current, such as inductance and capacitance.[] These properties however can become important when circuitry is subjected to transients, such as when first energised.

Electric field

Main article: Electric field. See also: Electrostatics.

The concept of the electric field was introduced by Michael Faraday. An electric field is created by a charged body in the space that surrounds it, and results in a force exerted on any other charges placed within the field. The electric field acts between two charges in a similar manner to the way that the gravitational field acts between two masses, and like it, extends towards infinity and shows an inverse square relationship with distance.[] However, there is an important difference. Gravity always acts in attraction, drawing two masses together, while the electric field can result in either attraction or repulsion. Since large bodies such as planets generally carry no net charge, the electric field at a distance is usually zero. Thus gravity is the dominant force at distance in the universe, despite being much weaker.[]

An electric field generally varies in space,[29] and its strength at any one point is defined as the force (per unit charge) that would be felt by a stationary, negligible charge if placed at that point.[] The conceptual charge, termed a 'test charge', must be vanishingly small to prevent its own electric field disturbing the main field and must also be stationary to prevent the effect of magnetic fields. As the electric field is defined in terms of force, and force is a vector, so it follows that an electric field is also a vector, having both magnitude and direction. Specifically, it is a vector field.[]

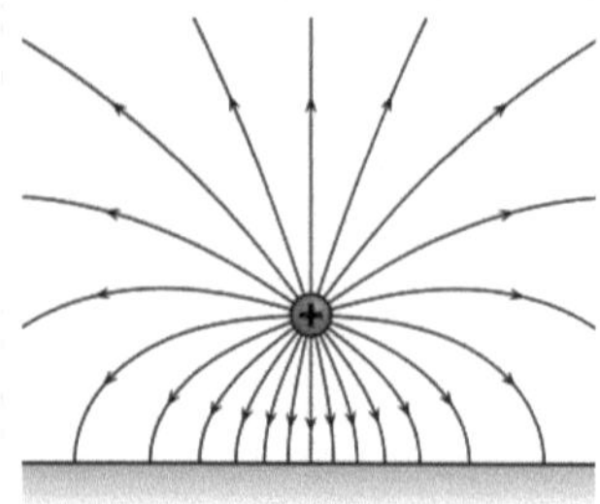

Field lines emanating from a positive charge above a plane conductor

The study of electric fields created by stationary charges is called electrostatics. The field may be visualised by a set of imaginary lines whose direction at any point is the same as that of the field. This concept was introduced by Faraday,[30] whose term 'lines of force' still sometimes sees use. The field lines are the paths that a point positive charge would seek to make as it was forced to move within the field; they are however an imaginary concept with no physical existence, and the field permeates all the intervening space between the lines.[30] Field lines emanating from stationary charges have several key properties: first, that they originate at positive charges and terminate at negative charges; second, that they must enter any good conductor at right angles, and third, that they may never cross nor close in on themselves.[31]

A hollow conducting body carries all its charge on its outer surface. The field is therefore zero at all places inside the body.[32] This is the operating principal of the Faraday cage, a conducting metal shell which isolates its interior from outside electrical effects.

The principles of electrostatics are important when designing items of high-voltage equipment. There is a finite limit to the electric field strength that may be withstood by any medium. Beyond this point, electrical breakdown occurs and an electric arc causes flashover between the charged parts. Air, for example, tends to arc across small gaps at electric field strengths which exceed 30 kV per centimetre. Over larger gaps, its breakdown strength is weaker, perhaps 1 kV per centimetre.[] The most visible natural occurrence of this is lightning, caused when charge becomes separated in the clouds by rising columns of air, and raises the electric field in the air to greater than it can withstand. The voltage of a large lightning cloud may be as high as 100 MV and have discharge energies as great as 250 kWh.[33]

The field strength is greatly affected by nearby conducting objects, and it is particularly intense when it is forced to curve around sharply pointed objects. This principle is exploited in the lightning conductor, the sharp spike of which acts to encourage the lightning stroke to develop there, rather than to the building it serves to protect.[34]

Electric potential

Main article: Electric potential. See also: Voltage, Battery (electricity)

The concept of electric potential is closely linked to that of the electric field. A small charge placed within an electric field experiences a force, and to have brought that charge to that point against the force requires work. The electric potential at any point is defined as the energy required to bring a unit test charge from an infinite distance slowly to that point. It is usually measured in volts, and one volt is the potential for which one joule of work must be expended to bring a charge of one coulomb from infinity.[] This definition of potential, while formal, has little practical application, and a more useful concept is that of electric potential difference, and is the energy required to move a unit charge between two specified points. An electric field has the special property that it is *conservative*, which means that the path taken by the test charge is irrelevant: all paths between two specified points expend the same energy, and thus a unique value for potential difference may be stated.[] The volt is so strongly identified as the unit of choice for measurement and description of electric potential difference that the term voltage sees greater everyday usage.

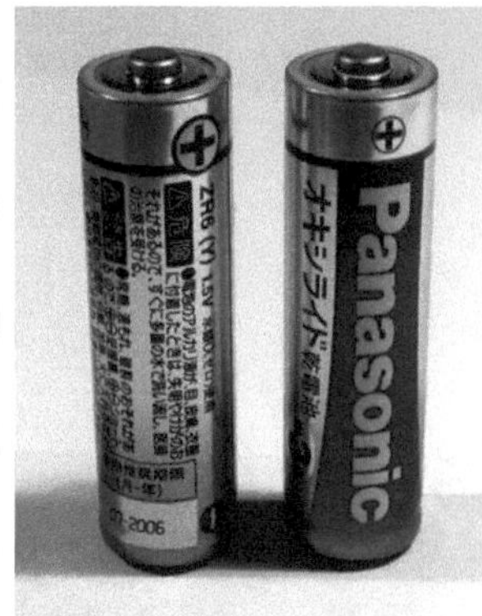

A pair of AA cells. The + sign indicates the polarity of the potential difference between the battery terminals.

For practical purposes, it is useful to define a common reference point to which potentials may be expressed and compared. While this could be at infinity, a much more useful reference is the Earth itself, which is assumed to be at the same potential everywhere. This reference point naturally takes the name earth or ground. Earth is assumed to be an infinite source of equal amounts of positive and negative charge, and is therefore electrically uncharged—and unchargeable.[35]

Electric potential is a scalar quantity, that is, it has only magnitude and not direction. It may be viewed as analogous to height: just as a released object will fall through a difference in heights caused by a gravitational field, so a charge will 'fall' across the voltage caused by an electric field.[36] As relief maps show contour lines marking points of equal height, a set of lines marking points of equal potential (known as equipotentials) may be drawn around an electrostatically charged object. The equipotentials cross all lines of force at right angles. They must also lie parallel to a conductor's surface, otherwise this would produce a force that will move the charge carriers to even the potential of the surface.

The electric field was formally defined as the force exerted per unit charge, but the concept of potential allows for a more useful and equivalent definition: the electric field is the local gradient of the electric potential. Usually expressed in volts per metre, the vector direction of the field is the line of greatest slope of potential, and where the equipotentials lie closest together.[]

Electromagnets

Magnetic field circles around a current

Ørsted's discovery in 1821 that a magnetic field existed around all sides of a wire carrying an electric current indicated that there was a direct relationship between electricity and magnetism. Moreover, the interaction seemed different from gravitational and electrostatic forces, the two forces of nature then known. The force on the compass needle did not direct it to or away from the current-carrying wire, but acted at right angles to it.[27] Ørsted's slightly obscure words were that "the electric conflict acts in a revolving manner." The force also depended on the direction of the current, for if the flow was reversed, then the force did too.[37]

Ørsted did not fully understand his discovery, but he observed the effect was reciprocal: a current exerts a force on a magnet, and a magnetic field exerts a force on a current. The phenomenon was further investigated by Ampère, who discovered that two parallel current-carrying wires exerted a force upon each other: two wires conducting currents in the same direction are attracted to each other, while wires containing currents in opposite directions are forced apart.[38] The interaction is mediated by the magnetic field each current produces and forms the basis for the international definition of the ampere.[38]

This relationship between magnetic fields and currents is extremely important, for it led to Michael Faraday's invention of the electric motor in 1821. Faraday's homopolar motor consisted of a permanent magnet sitting in a pool of mercury. A current was allowed through a wire suspended from a pivot above the magnet and dipped into the mercury. The magnet exerted a tangential force on the wire, making it circle around the magnet for as long as the current was maintained.[]

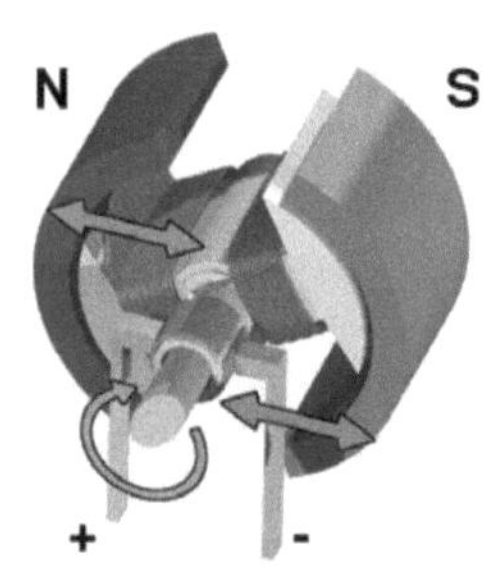

The electric motor exploits an important effect of electromagnetism: a current through a magnetic field experiences a force at right angles to both the field and current

Experimentation by Faraday in 1831 revealed that a wire moving perpendicular to a magnetic field developed a potential difference between its ends. Further analysis of this process, known as electromagnetic induction, enabled him to state the principle, now known as Faraday's law of induction, that the potential difference induced in a closed circuit is proportional to the rate of change of magnetic flux through the loop. Exploitation of this discovery enabled him to invent the first electrical generator in 1831, in which he converted the mechanical energy of a rotating copper disc to electrical energy.[] Faraday's disc was inefficient and of no use as a practical generator, but it showed the possibility of generating electric power using magnetism, a possibility that would be taken up by those that followed on from his work.

Faraday's and Ampère's work showed that a time-varying magnetic field acted as a source of an electric field, and a time-varying electric field was a source of a magnetic field. Thus, when either field is changing in time, then a field of the other is necessarily induced.[] Such a phenomenon has the properties of a wave, and is naturally referred to as an electromagnetic wave. Electromagnetic waves were analysed theoretically by James Clerk Maxwell in 1864. Maxwell developed a set of equations that could unambiguously describe the interrelationship between electric field, magnetic field, electric charge, and electric current. He could moreover prove that such a wave would necessarily travel at the speed of light, and thus light itself was a form of electromagnetic radiation. Maxwell's Laws, which unify light, fields, and charge are one of the great milestones of theoretical physics.[]

Electric circuits

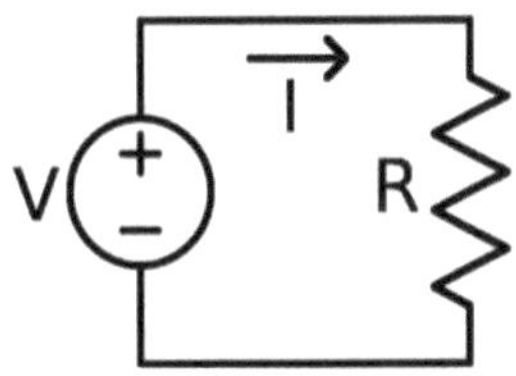

A basic electric circuit. The voltage source *V* on the left drives a current *I* around the circuit, delivering electrical energy into the resistor *R*. From the resistor, the current returns to the source, completing the circuit.

An electric circuit is an interconnection of electric components such that electric charge is made to flow along a closed path (a circuit), usually to perform some useful task.

The components in an electric circuit can take many forms, which can include elements such as resistors, capacitors, switches, transformers and electronics. Electronic circuits contain active components, usually semiconductors, and typically exhibit non-linear behaviour, requiring complex analysis. The simplest electric components are those that are termed passive and linear: while they may temporarily store energy, they contain no sources of it, and exhibit linear responses to stimuli.[]

The resistor is perhaps the simplest of passive circuit elements: as its name suggests, it resists the current through it, dissipating its energy as heat. The resistance is a consequence of the motion of charge through a conductor: in metals, for example, resistance is primarily due to collisions between electrons and ions. Ohm's law is a basic law of circuit theory, stating that the current passing through a resistance is directly proportional to the potential difference across it. The resistance of most materials is relatively constant over a range of temperatures and currents; materials under these conditions are known as 'ohmic'. The ohm, the unit of resistance, was named in honour of Georg Ohm, and is symbolised by the Greek letter Ω. 1 Ω is the resistance that will produce a potential difference of one volt in response to a current of one amp.[]

The capacitor is a development of the Leyden jar and is a device that can store charge, and thereby storing electrical energy in the resulting field. It consists of two conducting plates separated by a thin insulating dielectric layer; in practice, thin metal foils are coiled together, increasing the surface area per unit volume and therefore the capacitance. The unit of capacitance is the farad, named after Michael Faraday, and given the symbol *F*: one farad is the capacitance that develops a potential difference of one volt when it stores a charge of one coulomb. A capacitor connected to a voltage supply initially causes a current as it accumulates charge; this current will however decay in time as the capacitor fills, eventually falling to zero. A capacitor will therefore not permit a steady state current, but instead blocks it.[]

The inductor is a conductor, usually a coil of wire, that stores energy in a magnetic field in response to the current through it. When the current changes, the magnetic field does too, inducing a voltage between the ends of the conductor. The induced voltage is proportional to the time rate of change of the current. The constant of proportionality is termed the inductance. The unit of inductance is the henry, named after Joseph Henry, a contemporary of Faraday. One henry is the inductance that will induce a potential difference of one volt if the current through it changes at a rate of one ampere per second.[] The inductor's behaviour is in some regards converse to that of the capacitor: it will freely allow an unchanging current, but opposes a rapidly changing one.

Production and uses

Generation and transmission

Main article: Electricity generation. See also: Electric power transmission and Mains power around the world.

Early 20th-century alternator made in Budapest, Hungary, in the power generating hall of a hydroelectric station (photograph by Prokudin-Gorsky, 1905–1915).

Thales' experiments with amber rods were the first studies into the production of electrical energy. While this method, now known as the triboelectric effect, can lift light objects and generate sparks, it is extremely inefficient.[] It was not until the invention of the voltaic pile in the eighteenth century that a viable source of electricity became available. The voltaic pile, and its modern descendant, the electrical battery, store energy chemically and make it available on demand in the form of electrical energy.[] The battery is a versatile and very common power source which is ideally suited to many applications, but its energy storage is finite, and once discharged it must be disposed of or recharged. For large electrical demands electrical energy must be generated and transmitted continuously over conductive transmission lines.

Electrical power is usually generated by electro-mechanical generators driven by steam produced from fossil fuel combustion, or the heat released from nuclear reactions; or from other sources such as kinetic energy extracted from wind or flowing water. The modern steam turbine invented by Sir Charles Parsons in 1884 today generates about 80 percent of the electric power in the world using a variety of heat sources. Such generators bear no resemblance to Faraday's homopolar disc generator of 1831, but they still rely on his electromagnetic principle that a conductor linking a changing magnetic field induces a potential difference across its ends.[39] The invention in the late nineteenth century of the transformer meant that electrical power could be transmitted more efficiently at a higher voltage but lower current. Efficient electrical transmission meant in turn that electricity could be generated at centralised power stations, where it benefited from economies of scale, and then be despatched relatively long distances to where it was needed.[] [40]

Wind power is of increasing importance in many countries

Since electrical energy cannot easily be stored in quantities large enough to meet demands on a national scale, at all times exactly as much must be produced as is required.[] This requires electricity utilities to make careful predictions of their electrical loads, and maintain constant co-ordination with their power stations. A certain amount of generation must always be held in reserve to cushion an electrical grid against inevitable disturbances and losses.

Demand for electricity grows with great rapidity as a nation modernises and its economy develops. The United States showed a 12% increase in demand during each year of the first three decades of the twentieth century,[41] a rate of growth that is now being experienced by emerging economies such as those of India or China.[42] [43] Historically, the growth rate for electricity demand has outstripped that for other forms of energy.[44]

Environmental concerns with electricity generation have led to an increased focus on generation from renewable sources, in particular from wind and hydropower. While debate can be expected to continue over the environmental impact of different means of electricity production, its final form is relatively clean.[45]

Uses

The use of electricity gives a very convenient way to transfer energy, and because of this it has been adapted to a huge, and growing, number of uses.[46] The invention of a practical incandescent light bulb in the 1870s led to lighting becoming one of the first publicly available applications of electrical power. Although electrification brought with it its own dangers, replacing the naked flames of gas lighting greatly reduced fire hazards within homes and factories.[47] Public utilities were set up in many cities targeting the burgeoning market for electrical lighting.

The light bulb, an early application of electricity, operates by Joule heating: the passage of current through resistance generating heat

The Joule heating effect employed in the light bulb also sees more direct use in electric heating. While this is versatile and controllable, it can be seen as wasteful, since most electrical generation has already required the production of heat at a power station.[48] A number of countries, such as Denmark, have issued legislation restricting or banning the use of electric heating in new buildings.[49] Electricity is however a highly practical energy source for refrigeration,[50] with air conditioning representing a growing sector for electricity demand, the effects of which electricity utilities are increasingly obliged to accommodate.[51]

Electricity is used within telecommunications, and indeed the electrical telegraph, demonstrated commercially in 1837 by Cooke and Wheatstone, was one of its earliest applications. With the construction of first intercontinental, and then transatlantic, telegraph systems in the 1860s, electricity had enabled communications in minutes across the globe. Optical fibre and satellite communication technology have taken a share of the market for communications systems, but electricity can be expected to remain an essential part of the process.

The effects of electromagnetism are most visibly employed in the electric motor, which provides a clean and efficient means of motive power. A stationary motor such as a winch is easily provided with a supply of power, but a motor that moves with its application, such as an electric vehicle, is obliged to either carry along a power source such as a battery, or to collect current from a sliding contact such as a pantograph, placing restrictions on its range or performance.

Electronic devices make use of the transistor, perhaps one of the most important inventions of the twentieth century,[52] and a fundamental building block of all modern circuitry. A modern integrated circuit may contain several billion miniaturised transistors in a region only a few centimetres square.[53]

Electricity is also used to fuel public transportation, including electric buses and trains. [54]

Two 1 New York City Subway Trains, running electrically.

Electricity and the natural world

Physiological effects

A voltage applied to a human body causes an electric current through the tissues, and although the relationship is non-linear, the greater the voltage, the greater the current.[] The threshold for perception varies with the supply frequency and with the path of the current, but is about 0.1 mA to 1 mA for mains-frequency electricity, though a current as low as a microamp can be detected as an electrovibration effect under certain conditions.[55] If the current is sufficiently high, it will cause muscle contraction, fibrillation of the heart, and tissue burns.[] The lack of any visible sign that a conductor is electrified makes electricity a particular hazard. The pain caused by an electric shock can be intense, leading electricity at times to be employed as a method of torture. Death caused by an electric shock is referred to as electrocution. Electrocution is still the means of judicial execution in some jurisdictions, though its use has become rarer in recent times.[56]

Electrical phenomena in nature

Electricity is not a human invention, and may be observed in several forms in nature, a prominent manifestation of which is lightning. Many interactions familiar at the macroscopic level, such as touch, friction or chemical bonding, are due to interactions between electric fields on the atomic scale. The Earth's magnetic field is thought to arise from a natural dynamo of circulating currents in the planet's core.[57] Certain crystals, such as quartz, or even sugar, generate a potential difference across their faces when subjected to external pressure.[] This phenomenon is known as piezoelectricity, from the Greek *piezein* (πιέζειν), meaning to press, and was discovered in 1880 by Pierre and Jacques Curie. The effect is reciprocal, and when a piezoelectric material is subjected to an electric field, a small change in physical dimensions takes place.[]

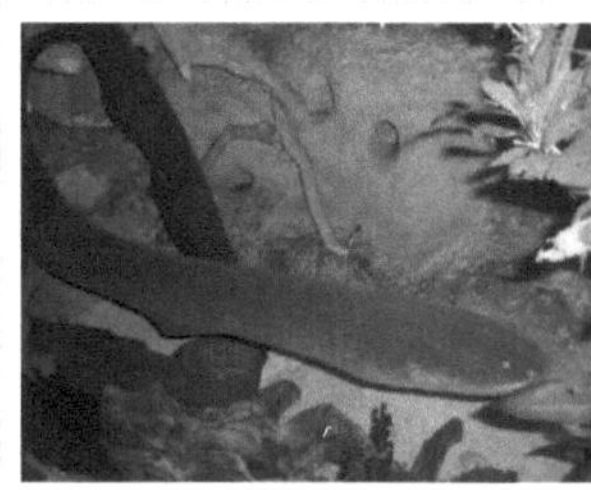

The electric eel, *Electrophorus electricus*

Some organisms, such as sharks, are able to detect and respond to changes in electric fields, an ability known as electroreception,[] while others, termed electrogenic, are able to generate voltages themselves to serve as a predatory or defensive weapon.[] The order Gymnotiformes, of which the best known example is the electric eel, detect or stun their prey via high voltages generated from modified muscle cells called electrocytes.[] [3] All animals transmit information along their cell membranes with voltage pulses called action potentials, whose functions include communication by the nervous system between neurons and muscles.[] An electric shock stimulates this system, and causes muscles to contract.[58] Action potentials are also responsible for coordinating activities in certain plants.[]

Cultural perception

In the 19th and early 20th century, electricity was not part of the everyday life of many people, even in the industrialised Western world. The popular culture of the time accordingly often depicts it as a mysterious, quasi-magical force that can slay the living, revive the dead or otherwise bend the laws of nature.[] This attitude began with the 1771 experiments of Luigi Galvani in which the legs of dead frogs were shown to twitch on application of animal electricity. "Revitalization" or resuscitation of apparently dead or drowned persons was reported in the medical literature shortly after Galvani's work. These results were known to Mary Shelley when she authored *Frankenstein* (1819), although she does not name the method of revitalization of the monster. The revitalization of monsters with electricity later became a stock theme in horror films.

As the public familiarity with electricity as the lifeblood of the Second Industrial Revolution grew, its wielders were more often cast in a positive light,[59] such as the workers who "finger death at their gloves' end as they piece and repiece the living wires" in Rudyard Kipling's 1907 poem *Sons of Martha*.[59] Electrically powered vehicles of every sort featured large in adventure stories such as those of Jules Verne and the *Tom Swift* books.[59] The masters of electricity, whether fictional or real—including scientists such as Thomas Edison, Charles Steinmetz or Nikola Tesla—were popularly conceived of as having wizard-like powers.[59]

With electricity ceasing to be a novelty and becoming a necessity of everyday life in the later half of the 20th century, it required particular attention by popular culture only when it *stops* flowing,[59] an event that usually signals disaster.[59] The people who *keep* it flowing, such as the nameless hero of Jimmy Webb's song "Wichita Lineman" (1968),[59] are still often cast as heroic, wizard-like figures.[59]

See also

- Ampère's circuital law, connects the direction of an electric current and its associated magnetic currents.
- Electric potential energy, the potential energy of a system of charges
- Electricity market, the sale of electrical energy
- Electrical phenomena, observable events which illuminate the physical principles of electricity
- Electric power, the rate at which electrical energy is transferred
- Electronics, the study of the movement of charge through certain materials and devices
- Hydraulic analogy, an analogy between the flow of water and electric current
- Mains electricity, the AC electric power supply
- Mains electricity by country, includes a list of countries and territories, with the plugs, voltages and frequencies they use

Notes

Footnotes

a. the New Latin *ēlectricus*, "amber-like", came from the classical Latin *electrum*, itself coming from the Greek ἤλεκτρον, (elektron), meaning amber

Citations

[4] *The Encyclopedia Americana; a library of universal knowledge* (1918), New York: Encyclopedia Americana Corp

[9] It is uncertain if Franklin personally carried out this experiment, but it is popularly attributed to him.

[11] Berkson, William (1974) Fields of force: the development of a world view from Faraday to Einstein (http://books.google.com/books?id=hMc9AAAAIAAJ&pg=PA148&dq=maxwell+on+physical+lines+of+force#v=onepage&q=maxwell on physical lines of force&f=false) p.148. Routledge, 1974

[13] Sears, Francis W., Mark W. Zemansky and Hugh D. Young (1983), *University Physics*, Sixth Edition, Addison-Wesley, pp. 843–4. ISBN 0-201-07195-9.

[15] "Solid state" (http://encyclopedia2.thefreedictionary.com/solid+state), *The Free Dictionary*

[16] John Sydney Blakemore, *Solid state physics*, pp.1-3, Cambridge University Press, 1985 ISBN 0-521-31391-0.

[17] Richard C. Jaeger, Travis N. Blalock, *Microelectronic circuit design*, pp.46-47, McGraw-Hill Professional, 2003 ISBN 0-07-250503-6.

[19] "The repulsive force between two small spheres charged with the same type of electricity is inversely proportional to the square of the distance between the centres of the two spheres." Charles-Augustin de Coulomb, *Histoire de l'Academie Royal des Sciences*, Paris 1785.

[23] . The Q originally stood for 'quantity of electricity', the term 'electricity' now more commonly expressed as 'charge'.

[27] Accounts differ as to whether this was before, during, or after a lecture.

[29] Almost all electric fields vary in space. An exception is the electric field surrounding a planar conductor of infinite extent, the field of which is uniform.

[59] Van Riper, op.cit., p. 71.

References

- Benjamin, P. (1898). A history of electricity (The intellectual rise in electricity) from antiquity to the days of Benjamin Franklin (http://books.google.com/books?id=VLsKAAAAIAAJ). New York: J. Wiley & Sons.

External links

- "One-Hundred Years of Electricity", May 1931, Popular Mechanics (http://books.google.com/books?id=n-MDAAAAMBAJ&pg=PA772&dq=Popular+Mechanics+1931+curtiss#v=onepage&q&f=true)
- Illustrated view of how an American home's electrical system works (http://www.hometips.com/hyhw/electrical/electric.html)
- Electricity around the world (http://users.pandora.be/worldstandards/electricity.htm)
- Electricity Misconceptions (http://amasci.com/miscon/elect.html)
- Electricity and Magnetism (http://www.micro.magnet.fsu.edu/electromag/java/diode/index.html)
- Understanding Electricity and Electronics in about 10 Minutes (http://steverose.com/Articles/UnderstandingBasicElectri.html)
- World Bank report on Water, Electricity and Utility subsidies (http://water.worldbank.org/water/publications/water-electricity-and-poor-who-benefits-utility-subsidies/)

Cogeneration

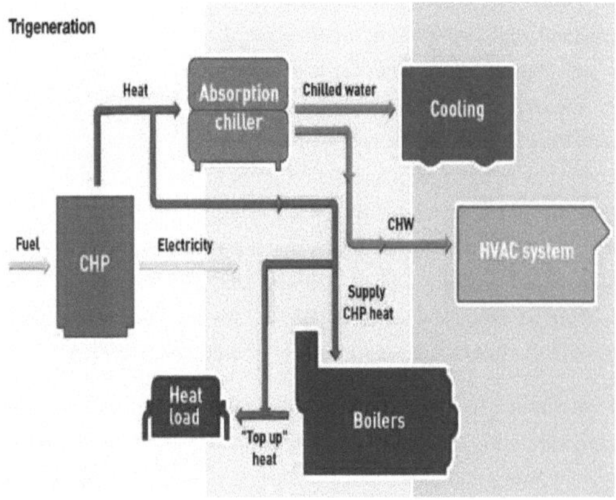

Trigeneration cycle

Cogeneration or **combined heat and power** (**CHP**) is the use of a heat engine[1] or power station to simultaneously generate electricity and useful heat. **Trigeneration** or **combined cooling, heat and power** (**CCHP**) refers to the simultaneous generation of electricity and useful heating and cooling from the combustion of a fuel or a solar heat collector. A plant producing electricity, heat and cold is called a trigeneration[2] or **polygeneration** plant.

Cogeneration is a thermodynamically efficient use of fuel. In separate production of electricity, some energy must be discarded as waste heat, but in cogeneration this thermal energy is put to use. All thermal power plants emit heat during electricity generation, which can be released into the natural environment through cooling towers, flue gas, or by other means. In contrast, CHP captures some or all of the by-product for heating, either very close to the plant, or—especially in Scandinavia and Eastern Europe—as hot water for district heating with temperatures ranging from approximately 80 to 130 °C. This is also called **combined heat and power district heating** (**CHPDH**). Small CHP plants are an example of decentralized energy.[3] By-product heat at moderate temperatures (100–180 °C, 212–356 °F) can also be used in absorption refrigerators for cooling.

The supply of high-temperature heat first drives a gas or steam turbine-powered generator and the resulting low-temperature waste heat is then used for water or space heating as described in cogeneration. Trigeneration differs from cogeneration in that the waste heat is used for both heating and cooling, typically in an absorption refrigerator. CCHP systems can attain higher overall efficiencies than cogeneration or traditional power plants. In the United States, the application of trigeneration in buildings is called **building cooling, heating and power** (**BCHP**). Heating and cooling output may operate concurrently or alternately depending on need and system construction.

Cogeneration was practiced in some of the earliest installations of electrical generation. Before central stations distributed power, industries generating their own power used exhaust steam for process heating. Large office and apartment buildings, hotels and stores commonly generated their own power and used waste steam for building heat. Due to the high cost of early purchased power, these CHP operations continued for many years after utility electricity became available.[4] Cogeneration is still common in pulp and paper mills, refineries and chemical plants.

In the United States, Consolidated Edison distributes 66 billion kilograms of 350 °F (180 °C) steam each year through its seven cogeneration plants to 100,000 buildings in Manhattan—the biggest steam district in the United States. The peak delivery is 10 million pounds per hour (or approximately 2.5 GW).[5] [6] Other major cogeneration companies in the United States include Recycled Energy Development,[7] and leading advocates include Tom Casten and Amory Lovins.

Overview

Masnedø CHP power station in Denmark. This station burns straw as fuel. The adjacent greenhouses are heated by district heating from the plant.

Thermal power plants (including those that use fissile elements or burn coal, petroleum, or natural gas), and heat engines in general, do not convert all of their thermal energy into electricity. In most heat engines, a bit more than half is lost as excess heat (see: Second law of thermodynamics and Carnot's theorem). By capturing the excess heat, CHP uses heat that would be wasted in a conventional power plant, potentially reaching an efficiency of up to 80%,[8] for the best conventional plants. This means that less fuel needs to be consumed to produce the same amount of useful energy.

Steam turbines for cogeneration are designed for *extraction* of steam at lower pressures after it has passed through a number of turbine stages, or they may be designed for final exhaust at *back pressure* (non-condensing), or both.[] A typical power generation turbine in a paper mill may have extraction pressures of 160 psig (1.103 MPa) and 60 psig (0.41 MPa). A typical back pressure may be 60 psig (0.41 MPa). In practice these pressures are custom designed for each facility. The extracted or exhaust steam is used for process heating, such as drying paper, evaporation, heat for chemical reactions or distillation. Steam at ordinary process heating conditions still has a considerable amount of enthalpy that could be used for power generation, so cogeneration has lost opportunity cost. Conversely, simply generating steam at process pressure instead of high enough pressure to generate power at the top end also has lost opportunity cost. (See: Steam turbine#Steam supply and exhaust conditions) The capital and operating cost of high pressure boilers, turbines and generators are substantial, and this equipment is normally operated continuously, which usually limits self generated power to large scale operations.

A cogeneration plant in Metz, France. The 45MW boiler uses waste wood biomass as energy source, and provides electricity and heat for 30,000 dwellings.

Some tri-cycle plants have used a combined cycle in which several thermodynamic cycles produced electricity, then a heating system was used as a condenser of the power plant's bottoming cycle. For example, the RU-25 MHD generator in Moscow heated a boiler for a conventional steam powerplant, whose condensate was then used for space heat. A more modern system might use a gas turbine powered by natural gas, whose exhaust powers a steam plant, whose condensate provides heat. Tri-cycle plants can have thermal efficiencies above 80%.

The viability of CHP (sometimes termed utilisation factor), especially in smaller CHP installations, depends on a good baseload of operation, both in terms of an on-site (or near site) electrical demand and heat demand. In practice, an exact match between the heat and electricity needs rarely exists. A CHP plant can either meet the need for heat (*heat driven operation*) or be run as a power plant with some use of its waste heat, the latter being less advantageous in terms of its utilisation factor and thus its overall efficiency. The viability can be greatly increased where opportunities for Trigeneration exist. In such cases, the heat from the CHP plant is also used as a primary energy source to deliver cooling by means of an absorption chiller.

CHP is most efficient when heat can be used on-site or very close to it. Overall efficiency is reduced when the heat must be transported over longer distances. This requires heavily insulated pipes, which are expensive and inefficient; whereas electricity can be transmitted along a comparatively simple wire, and over much longer distances for the same energy loss.

A car engine becomes a CHP plant in winter when the reject heat is useful for warming the interior of the vehicle. The example illustrates the point that deployment of CHP depends on heat uses in the vicinity of the heat engine.

Cogeneration plants are commonly found in district heating systems of cities, hospitals, prisons, oil refineries, paper mills, wastewater treatment plants, thermal enhanced oil recovery wells and industrial plants with large heating needs.

Thermally enhanced oil recovery (TEOR) plants often produce a substantial amount of excess electricity. After generating electricity, these plants pump leftover steam into heavy oil wells so that the oil will flow more easily, increasing production. TEOR cogeneration plants in Kern County, California produce so much electricity that it cannot all be used locally and is transmitted to Los Angeles.

CHP is one of the most cost-efficient methods of reducing carbon emissions from heating systems in cold climates.[9]

Utility pressures versus self generating industrial

Industrial cogeneration plants normally operate at much lower boiler pressures than utilities. Among the reasons are: 1) Cogeneration plants face possible contamination of returned condensate. Because boiler feed water from cogeneration plants has much lower return rates than 100% condensing power plants, industries usually have to treat proportionately more boiler make up water. Boiler feed water must be completely oxygen free and de-mineralized, and the higher the pressure the more critical the level of purity of the feed water.[] 2) Utilities are typically larger scale power than industry, which helps offset the higher capital costs of high pressure. 3) Utilities are less likely to have sharp load swings than industrial operations, which deal with shutting down or starting up units that may represent a significant percent of either steam or power demand.

Comparison with a heat pump

A heat pump may be compared with a CHP unit, in that for a condensing steam plant, as it switches to produced heat, then electrical power is lost or becomes unavailable, just as the power used in a heat pump becomes unavailable. Typically for every unit of power lost, then about 6 units of heat are made available at about 90°C. Thus CHP has an effective Coefficient of Performance (COP) compared to a heat pump of 6.[10] It is noteworthy that the unit for the CHP is lost at the high voltage network and therefore incurs no losses, whereas the heat pump unit is lost at the low voltage part of the network and incurs on average a 6% loss. Because the losses are proportional to the square of the current, during peak periods losses are much higher than this and it is likely that widespread i.e. city wide application of heat pumps would cause overloading of the distribution and transmission grids unless they are substantially reinforced.

It is also possible to run a heat driven operation combined with a heat pump, where the excess electricity (as heat demand is the defining factor on utilization) is used to drive a heat pump. As heat demand increases, more electricity is generated to drive the heat pump, with the waste heat also heating the heating fluid.

Thermal efficiency

Every heat engine is subject to the theoretical efficiency limits of the Carnot cycle. When the fuel is natural gas, a gas turbine following the Rankine cycle is typically used.[11] Mechanical energy from the turbine drives an electric generator. The low-grade (i.e. low temperature) waste heat rejected by the turbine is then applied to space heating or cooling or to industrial processes. Cooling is achieved by passing the waste heat to an absorption chiller.

Thermal efficiency in a trigeneration system is defined as:

Where:

= Thermal efficiency

= Total work output by all systems

= Total heat input into the system

Typical trigeneration models have losses as in any system. The energy distribution below is represented as a percent of total input energy:[12]

Electricity = 45%

Heat + Cooling = 40%

Heat Losses = 13%

Line Losses = 2%

Conventional central coal- or nuclear-powered power stations convert only about 33% of their input heat to electricity. The remaining 67% emerges from the turbines as low-grade waste heat with no significant local uses so it is usually rejected to the environment. These low conversion efficiencies strongly suggest that productive uses be found for this waste heat, and in some countries these plants do produce byproduct steam that can be sold to customers.

But if no practical uses can be found for the waste heat from a central power station, e.g., due to distance from potential customers, then moving generation to where the waste heat can find uses may be of great benefit. Even though the efficiency of a small distributed electrical generator may be lower than a large central power plant, the use of its waste heat for local heating and cooling can result in an overall use of the primary fuel supply as great as 80%. This provides substantial financial and environmental benefits.

Distributed generation

Trigeneration has its greatest benefits when scaled to fit buildings or complexes of buildings where electricity, heating and cooling are perpetually needed. Such installations include but are not limited to: data centers, manufacturing facilities, universities, hospitals, military complexes and colleges. Localized trigeneration has addition benefits as described by distributed generation. Redundancy of power in mission critical applications, lower power usage costs and the ability to sell electrical power back to the local utility are a few of the major benefits. Even for small buildings such as individual family homes trigeneration systems provide benefits over cogeneration because of increased energy utilization.[13]

Most industrial countries generate the majority of their electrical power needs in large centralized facilities with capacity for large electrical power output. These plants have excellent economies of scale, but usually transmit electricity long distances resulting in sizable losses, negatively affect the environment. Large power plants can use cogeneration or trigeneration systems only when sufficient need exists in immediate geographic vicinity for an industrial complex, additional power plant or a city. An example of cogeneration with trigeneration applications in a major city is the New York City steam system.

Types of plants

Topping cycle plants primarily produce electricity from a steam turbine. The exhausted steam is then condensed and the low temperature heat released from this condensation is utilized for e.g. district heating or water desalination.

Bottoming cycle plants produce high temperature heat for industrial processes, then a waste heat recovery boiler feeds an electrical plant. Bottoming cycle plants are only used when the industrial process requires very high temperatures such as furnaces for glass and metal manufacturing, so they are less common.

Large cogeneration systems provide heating water and power for an industrial site or an entire town. Common CHP plant types are:

- Gas turbine CHP plants using the waste heat in the flue gas of gas turbines. The fuel used is typically natural gas
- Gas engine CHP plants use a reciprocating gas engine which is generally more competitive than a gas turbine up to about 5 MW. The gaseous fuel used is normally natural gas. These plants are generally manufactured as fully packaged units that can be installed within a plantroom or external plant compound with simple connections to the site's gas supply and electrical distribution and heating systems. Typical large example see [14]
- Biofuel engine CHP plants use an adapted reciprocating gas engine or diesel engine, depending upon which biofuel is being used, and are otherwise very similar in design to a Gas engine CHP plant. The advantage of using a biofuel is one of reduced hydrocarbon fuel consumption and thus reduced carbon emissions. These plants are generally manufactured as fully packaged units that can be installed within a plantroom or external plant compound with simple connections to the site's electrical distribution and heating systems. Another variant is the wood gasifier CHP plant whereby a wood pellet or wood chip biofuel is gasified in a zero oxygen high temperature environment; the resulting gas is then used to power the gas engine. Typical smaller size biogas plant see [15]
- Combined cycle power plants adapted for CHP
- Steam turbine CHP plants that use the heating system as the steam condenser for the steam turbine.
- Molten-carbonate fuel cells and solid oxide fuel cells have a hot exhaust, very suitable for heating.
- Nuclear power plants can be fitted with taps after the turbines to provide steam to a heating system. With a heating system temperature of 95°C it is possible to extract about 10 MW heat for every MW electricity lost. With a temperature of 130°C the gain is slightly smaller, about 7 MW for every MWe lost.[16]

Smaller cogeneration units may use a reciprocating engine or Stirling engine. The heat is removed from the exhaust and radiator. The systems are popular in small sizes because small gas and diesel engines are less expensive than small gas- or oil-fired steam-electric plants.

Some cogeneration plants are fired by biomass,[17] or industrial and municipal waste (see incineration).

Some cogeneration plants combine gas and solar photovoltaic generation to further improve technical and environmental performance. [18]

Heat recovery steam generators

A heat recovery steam generator (HRSG) is a steam boiler that uses hot exhaust gases from the gas turbines or reciprocating engines in a CHP plant to heat up water and generate steam. The steam, in turn, drives a steam turbine or is used in industrial processes that require heat.

HRSGs used in the CHP industry are distinguished from conventional steam generators by the following main features:

- The HRSG is designed based upon the specific features of the gas turbine or reciprocating engine that it will be coupled to.
- Since the exhaust gas temperature is relatively low, heat transmission is accomplished mainly through convection.
- The exhaust gas velocity is limited by the need to keep head losses down. Thus, the transmission coefficient is low, which calls for a large heating surface area.

- Since the temperature difference between the hot gases and the fluid to be heated (steam or water) is low, and with the heat transmission coefficient being low as well, the evaporator and economizer are designed with plate fin heat exchangers.

MicroCHP

"Micro cogeneration" is a so-called distributed energy resource (DER). The installation is usually less than 5 kW_e in a house or small business. Instead of burning fuel to merely heat space or water, some of the energy is converted to electricity in addition to heat. This electricity can be used within the home or business or, if permitted by the grid management, sold back into the electric power grid. This recent development of small scale CHP systems has provided the opportunity for in-house power backup of residential-scale photovoltaic (PV) arrays.[19] The results of a recent study show that a PV+CHP hybrid system not only has the potential to radically reduce energy waste in the status quo electrical and heating systems, but it also enables the share of solar PV to be expanded by about a factor of five.[19] In some regions, in order to reduce waste from excess heat, an absorption chiller has been proposed to utilize the CHP-produced thermal energy for cooling of PV-CHP system. [20] These trigeneration+photovoltaic systems have the potential to save even more energy and further reduce emissions compared to conventional sources of power, heating and cooling. [21]

MiniCHP

Mini cogeneration is a so-called distributed energy resource (DER). The installation is usually more than 5 kW_e and less than 500 kW_e in a building or medium sized business. In this size range the viability or utilisation factor of the CHP plant is very important to consider since it will greatly affect the efficiency and cost effectiveness (payback) of the CHP plant. The utilisation factor is essentially the calculated hours of operation of the CHP plant expressed as a percentage of the total number of hours in a year. If less than 40% then the application of CHP is considered to be unviable. To be viable a good baseload for electrical demand and heat demand must exist. Such baseloads arise where building occupation or process activities are extended or continuous in operation. This typically includes for hospitals, prisons, manufacturing processes, swimming pools, airports, hotels, apartment blocks, etc.

Current (2007) Micro- and MiniCHP installations use five different technologies: microturbines, internal combustion engines, stirling engines, closed cycle steam engines and fuel cells. One author indicates that MicroCHP based on Stirling engines is the most cost effective of the so-called microgeneration technologies in abating carbon emissions;[22] however, advances in reciprocation engine technology are adding efficiency to CHP plant, particularly in the biogas field.[23] As both MiniCHP and CHP have been shown to reduce emissions [24] they could play a large role in the field of CO_2 reduction from buildings, where more than 14% of emissions can be saved using CHP in buildings.[25]

Costs

Typically, for a gas-fired plant the fully installed cost per kW electrical is around £400/kW, which is comparable with large central power stations.[15]

See also Cost of electricity by source

History

Cogeneration in Europe

A cogeneration thermal power plant in Ferrera Erbognone (PV), Italy

The EU has actively incorporated cogeneration into its energy policy via the CHP Directive. In September 2008 at a hearing of the European Parliament's Urban Lodgment Intergroup, Energy Commissioner Andris Piebalgs is quoted as saying, "security of supply really starts with energy efficiency."[26] Energy efficiency and cogeneration are recognized in the opening paragraphs of the European Union's Cogeneration Directive 2004/08/EC. This directive intends to support cogeneration and establish a method for calculating cogeneration abilities per country. The development of cogeneration has been very uneven over the years and has been dominated throughout the last decades by national circumstances.

As a whole, the European Union generates 11% of its electricity using cogeneration, saving Europe an estimated 35 Mtoe per annum a day.[27] However, there is large difference between Member States with variations of the energy savings between 2% and 60%. Europe has the three countries with the world's most intensive cogeneration economies: Denmark, the Netherlands and Finland.[28]

Other European countries are also making great efforts to increase efficiency. Germany reported that at present, over 50% of the country's total electricity demand could be provided through cogeneration. So far, Germany has set the target to double its electricity cogeneration from 12.5% of the country's electricity to 25% of the country's electricity by 2020 and has passed supporting legislation accordingly in "Federal Ministry of Economics and Technology, (BMWi), Germany, August 2007. The UK is also actively supporting combined heat and power. In light of UK's goal to achieve a 60% reduction in carbon dioxide emissions by 2050, the government has set the target to source at least 15% of its government electricity use from CHP by 2010.[29] Other UK measures to encourage CHP growth are financial incentives, grant support, a greater regulatory framework, and government leadership and partnership.

According to the IEA 2008 modeling of cogeneration expansion for the G8 countries, the expansion of cogeneration in France, Germany, Italy and the UK alone would effectively double the existing primary fuel savings by 2030. This would increase Europe's savings from today's 155.69 Twh to 465 Twh in 2030. It would also result in a 16% to 29% increase in each country's total cogenerated electricity by 2030.

Governments are being assisted in their CHP endeavors by organizations like COGEN Europe who serve as an information hub for the most recent updates within Europe's energy policy. COGEN is Europe's umbrella organization representing the interests of the cogeneration industry, users of the technology and promoting its benefits in the EU and the wider Europe. The association is supported by the key players in the industry including gas and electricity companies, ESCOs, equipment suppliers, consultancies, national promotion organisations, financial and other service companies.

Cogeneration in the United States

Perhaps the first modern use of energy recycling was done by Thomas Edison. His 1882 Pearl Street Station, the world's first commercial power plant, was a combined heat and power plant, producing both electricity and thermal energy while using waste heat to warm neighboring buildings.[30] Recycling allowed Edison's plant to achieve approximately 50 percent efficiency.

A 250 MW cogeneration plant in Cambridge, Massachusetts

By the early 1900s, regulations emerged to promote rural electrification through the construction of centralized plants managed by regional utilities. These regulations not only promoted electrification throughout the countryside, but they also discouraged decentralized power generation, such as cogeneration. As Recycled Energy Development CEO Sean Casten testified to Congress, they even went so far as to make it illegal for non-utilities to sell power.[31]

By 1978, Congress recognized that efficiency at central power plants had stagnated and sought to encourage improved efficiency with the Public Utility Regulatory Policies Act (PURPA), which encouraged utilities to buy power from other energy producers.

Percentage of U.S. energy produced by cogeneration

Cogeneration plants proliferated, soon producing about 8% of all energy in the United States.[] However, the bill left implementation and enforcement up to individual states, resulting in little or nothing being done in many parts of the country.

In 2008 Tom Casten, chairman of Recycled Energy Development, said that "We think we could make about 19 to 20 percent of U.S. electricity with heat that is currently thrown away by industry."[32]

The United States Department of Energy has an aggressive goal of having CHP comprise of 20% of generation capacity by the year 2030. Eight Clean Energy Application Centers[33] have been established across the nation whose mission is to develop the required technology application knowledge and educational infrastructure necessary to lead "clean energy" (combined heat and power, waste heat recovery and district energy) technologies as viable energy options and reduce any perceived risks associated with their implementation. The focus of the Application Centers is to provide an outreach and technology deployment program for end users, policy makers, utilities, and industry stakeholders.

Outside of the United States, energy recycling is more common. Denmark is probably the most active energy recycler, obtaining about 55% of its energy from cogeneration and waste heat recovery. Other large countries, including Germany, Russia, and India, also obtain a much higher share of their energy from decentralized sources.[] [32]

Applications in power generation systems

Non-renewable

Any of the following conventional power plants may be converted to a CCHP system:[34]

- Coal
- Microturbine
- Natural gas
- Nuclear power
- Oil
- Small gas turbine

Renewable

- Fuel cell
- Solar power—both solar thermal and photovoltaic
- Biomass

See also

- Air separation
- Carnot cycle
- CHP Directive
- Cost of electricity by source
- Distributed generation (more general term encompassing CHP)
- District heating
- Energy policy of the European Union
- Environmental impact of electricity generation
- Euroheat & Power
- Industrial gas
- Micro combined heat and power
- New York City steam system
- Rankine cycle
- Renewable energy in Australia
- Syngas

Further reading

- An engineering handbook widely used by those involved with various types of boilers. Contains numerous illustrations, graphs and useful formulas. (Not specific to cogeneration). *The link leads to an entire free e-Book of an early edition. For current practice a more modern edition is recommended.*

External links

- CHP in the United States:
  - Southeast CHP Application Center [35].
  - Gulf Coast (Texas, Louisiana & Oklahoma) [36].
  - Northwest [37].
  - Midwest [38].
  - Intermountain [39].

- Mid-Atlantic [40].
- Northeast [41].
- Pacific [42].
- CHP in Finland:
  - High cogeneration performance by innovative steam turbine for biomass-fired CHP plant in Iisalmi, Finland [43].
- CHP in Belgium :
  - Experimental study and modeling of a low temperature Rankine Cycle for small scale cogeneration [44]
- CHP in the UK :
  - CHP Focus - DECC's free resource for helping to develop combined heat and power [45]
  - BBC News: Power from the people [46]
  - M.I.T. algae reactor [47]
  - Future Challenges for CHP in the UK and Continental Europe, Feb 2010 by F.Star [48]
- Associations:
  - The UK District Energy Association - advocating the construction of local energy networks [49]
  - U.S. Clean Heat and Power Association [50]
  - COGEN Europe The European Association for the Promotion of Cogeneration [51]
  - U.K. Combined Heat and Power Association, promotes the wider use of combined heat and power and community heating. [52]
  - List of all CHP releated Associations and Companies [53]
- Briefings:
  - The Role of District Energy/Combined Heat and Power in Energy and Climate Policy Solutions [54] Environmental and Energy Study Institute Briefing
- Trigeneration
  - 3D model of installed commercial trigeneration system [55]
  - Absorption chiller animation [56]
  - European site on trigeneration [57]
  - Commercial absorption chiller for trigeneration [58]
  - Commercial trigeneration system [59]
  - Commercial trigeneration system [60]

References

[1] Cogeneration and Cogeneration Schematic (http://www.clarke-energy.com/chp-cogeneration/), www.clarke-energy.com, retrieved 26.11.11

[10] ScienceDirect.com - Energy Policy - Combined heat and power considered as a virtual steam cycle heat pump (http://www.sciencedirect.com/science/article/pii/S030142151100379X)

[13] A.H. Nosrat, L.G. Swan, J.M. Pearce, " Improved Performance of Hybrid Photovoltaic-Trigeneration Systems Over Photovoltaic-Cogen Systems Including Effects of Battery Storage (http://www.academia.edu/2337798/Improved_Performance_of_Hybrid_Photovoltaic-Trigeneration_Systems_Over_Photovoltaic-Cogen_Systems_Including_Effects_of_Battery_Storage)", *Energy* 49, pp. 366-374 (2013).DOI:10.1016/j.energy.2012.11.005

[15] 38% HHV Caterpillar Bio-gas Engine Fitted to Sewage Works | Claverton Group (http://www.claverton-energy.com/38-hhv-caterpillar-bio-gas-engine-fitted-to-long-reach-sewage-works.html)

[16] http://www.elforsk.se/nyhet/seminarie/Elforskdagen%20_10/webb_varme/d_welander.pdf [swedish]

[18] A.C. Oliveira, C. Afonso, J. Matos, S. Riffat, M. Nguyen and P. Doherty, " **A Combined Heat and Power System for Buildings driven by Solar Energy and Gas** (http://dx.doi.org/10.1016/S1359-4311(01)00110-7)", *Applied Thermal Engineering*, vol. **22**, Iss. 6, pp. 587-593 (2002).

[19] J. M. Pearce, "Expanding Photovoltaic Penetration with Residential Distributed Generation from Hybrid Solar Photovoltaic + Combined Heat and Power Systems", *Energy* **34**, pp. 1947-1954 (2009). (http://dx.doi.org/10.1016/j.energy.2009.08.012) Open access (http://hdl.

handle.net/1974/5307)
[20] A. Nosrat and J. M. Pearce, "Dispatch Strategy and Model for Hybrid Photovoltaic and Combined Heating, Cooling, and Power Systems", *Applied Energy* **88** (2011) 3270–3276. (http://dx.doi.org/10.1016/j.apenergy.2011.02.044) Open access (http://hdl.handle.net/1974/6439)
[21] A.H. Nosrat, L.G. Swan, J.M. Pearce, "Improved Performance of Hybrid Photovoltaic-Trigeneration Systems Over Photovoltaic-Cogen Systems Including Effects of Battery Storage", *Energy* **49**, pp. 366-374 (2013). DOI (http://dx.doi.org/10.1016/j.energy.2012.11.005), open access (http://www.academia.edu/2337798/Improved_Performance_of_Hybrid_Photovoltaic-Trigeneration_Systems_Over_Photovoltaic-Cogen_Systems_Including_Effects_of_Battery_Storage).
[22] What is microgeneration? Jeremy Harrison, Claverton Energy Group Conference, Bath, Oct 24th 2008 (http://www.claverton-energy.com/what-is-microgeneration.html)
[23] MiniCHP ranges and efficiencies Aug 15 2009 (http://www.alfagy.com/)
[24] Pehnt, M. (2008). Environmental impacts of distributed energy systems—The case of micro cogeneration. Environmental science & policy, 11(1), 25-37.
[25] http://alfagy.com/what-is-chp/133-kaarsberg-t-rfiskum-jromm-a-rosenfeld-j-koomey-and-wpteagan-1998-qcombined-heat-and-power-chp-or-cogeneration-for-saving-energy-and-carbon-in-coı html "Combined Heat and Power (CHP or Cogeneration) for Saving Energy and Carbon in Commercial Buildings."
[32] 'Recycling' Energy Seen Saving Companies Money (http://www.npr.org/templates/story/story.php?storyId=90714692). By David Schaper. May 22, 2008. Morning Edition. National Public Radio.
[33] Eight Clean Energy Application Centers (http://www.gulfcoastcleanenergy.org/WhatisCHP/StateInformation/OtherStates/tabid/1349/Default.aspx)
[35] http://www.chpcenterse.org
[36] http://www.gulfcoastcleanenergy.org/
[37] http://www.chpcenternw.org/
[38] http://www.chpcentermw.org/home.html
[39] http://www.intermountainchp.org/
[40] http://www.maceac.psu.edu/
[41] http://www.northeastchp.org/nac/index.cfm
[42] http://www.chpcenterpr.org/
[43] http://www.opet-chp.net/download/wp3/iisalmifinland.pdf
[44] http://www.labothap.ulg.ac.be/cmsms/Staff/QuoilinS/TFE_SQ010607.pdf
[45] http://chp.defra.gov.uk/cms/
[46] http://news.bbc.co.uk/2/hi/programmes/working_lunch/3231549.stm
[47] http://cogen.mit.edu/powermit/
[48] http://www.claverton-energy.com/future-challenges-for-chp-in-the-uk-and-continental-europe-f-starr-phd-fimmm-c-eng-feb-2010.html
[49] http://www.ukdea.org.uk/
[50] http://www.uschpa.org/
[51] http://www.cogeneurope.eu/
[52] http://www.chpa.co.uk/
[53] http://alfagy.com/index.php?option=com_weblinks&view=category&id=12&Itemid=44/
[54] http://www.eesi.org/042109_district_energy
[55] http://www.youtube.com/watch?v=o0L6dqH5qpI&NR=1
[56] http://www.youtube.com/watch?v=vHLKKd04hfk
[57] http://www.trigeneration.eu
[58] http://www.eco-maxchillers.com/common/Content.asp?PAGE=362&CONTENT=865
[59] http://www.2g-cenergy.com/trigen-more.html
[60] http://www.marathonengine.com/ecopower_principles.html

KS_150

KS 150 is a heavy water Gas Cooled Reactor (HWGCR) nuclear reactor design. A single example, **A-1**, was constructed at the Bohunice Nuclear Power Plant in Jaslovské Bohunice, Czechoslovakia. The power plant suffered a series of accidents, the worst being an accident on February 22, 1977 rated INES-4. Since 1979 the plant has been undergoing decommissioning.

History

The decision to build a nuclear power plant in Czechoslovakia was made in 1956. Construction of A-1 in Jaslovské Bohunice (western Slovakia) started in 1958 and took an unexpected 16 years. A-1 was commissioned on October 24, 1972.[]

The KS 150 reactor was designed in the Soviet Union and built entirely in Czechoslovakia, by Škoda Works. This type of reactor had never been deployed before. One advantage of the design was its ability to use unenriched uranium mined in Czechoslovakia, similar to a CANDU reactor.

Because of its experimental design the power plant suffered from accidents resulting in over 30 unplanned shutdowns. On January 5, 1976 two workers were killed due to a leak of carbon dioxide, which was used as a coolant. A 'technical' (mechanical?) failure occurred during refuelling and a fresh fuel assembly was shot off the reactor into the reactor's hall.[] The most serious accident from 1977 (see below) was rated INES-4. The damage could have been repaired with a large investment but on May 17, 1979 the government, dissatisfied with high costs, low performance and accidents, decided to decommission the plant. Plans to build the second reactor block A-2 were canceled.

The accidents were kept secret, although wild stories circulated among the public.

A1 Nuclear Power Plant was in operation for 19,261 hours altogether, it generated 1,464 GWh and supplied 916 GWh to the network. Maximum output achieved was 127 MW.[]

Decommissioning, decontamination and dismantling of the plant still continues and is expected to be completed in 2033.[1]

Technical details

KS 150 is a heavy water moderated, gas cooled reactor (HWGCR) able to refuel during operation.

Seventy metal uranium wires, each clad in a compound of magnesium and beryllium, are bundled together to form a fuel rod.

The reactor's pressure vessel is of 15 cm carbon steel in a cylindrical shape with diameter 5.1 m and height 20 m. Within the pressure vessel (in the active zone) is a cylindrical vessel of aluminium- magnesium-silicon alloy for the heavy water moderator.[2]

Fuel channels are vertical, each containing a single fuel rod cooled with circulating carbon dioxide. The core is in a pressurized vessel to allow refuelling during operation. The heavy water moderator is cooled in a separate circuit.

Carbon Dioxide gas used as a primary coolant is streamed around the fuel rods. After being heated by the rods it is piped to six steam generators. The resulting steam powers three turbogenerators.

- Fuel: unenriched metal uranium, 23.1 tonnes in the reactor.
- Core: diameter 3.56 m, height 4 m.
- Coolant gas on exit from reactor: pressure 5.4 MPa (~54 atm), temperature 426°C.
- Conversion efficiency: 18.5%.
- Moderating heavy water: temperature 65°C (Max/exit 90°C)
- Capacity: 143 MWe.

1977 accident

On February 22, 1977, during a fuel change, a combination of human mistakes and design problems caused the worst nuclear accident in Czechoslovakian history. Some fuel rods were being replaced while the reactor was active in a standard procedure. In this instance however humidity absorbers covering the rods were not removed, causing local overheating of the fuel (since transmission of heat to the coolant gas was reduced). The active zone was damaged, heavy water got in contact with the coolant and both primary and secondary circuits were contaminated.

The accident was rated as level 4 on International Nuclear Event Scale (in comparison, the Three Mile Island accident was rated level 5).

25% of the fuel elements in a heavy water moderated carbon dioxide cooled 100 MW(e) power reactor were damaged due to operator error. The operators failed to remove silica gel packs from a new fuel element. The silica gel was used to keep the unused fuel dry during storage and transport. The silica gel packs blocked the flow of the coolant resulting in overheating of the fuel and the pressure channel holding it. As a result of overheating the heavy water leaked into the part of the reactor (the gas circuit) where the fuel elements are accommodated, the fuel cladding was subject to corrosion and a considerable amount of radioactivity leaked into the primary cooling circuit (CO2 gas). Through leaks in the steam boilers (similar basic design to a MAGNOX or AGR plant) some parts of the secondary circuit became contaminated.[3]

References

[1] http://www.javys.sk/en/index.php?page=vyradovanie-jadrovoenergetickych-zariadeni/vyradovanie-jadrovej-elektrarne-a1/projekt-vyradovania Decommissioning project, retrieved 2010 Nov 18

[3] Page 300, Radioactivity, Ionizing Radiation and Nuclear Energy, Jiří Hála and James D. Navratil, Published by Konvoj (Brno) 2003, ISBN 80-7302-053-X

External links

- Detailed schematics of the KS-150 reactor (http://www.javys.sk/en/index.php?page=vyradovanie-jadrovoenergetickych-zariadeni/vyradovanie-jadrovej-elektrarne-a1/technologia/reaktor-ks-150)
- Detailed schematics of the A-1 nuclear power plant (http://www.javys.sk/en/index.php?page=vyradovanie-jadrovoenergetickych-zariadeni/vyradovanie-jadrovej-elektrarne-a1/technologia)
- Short description, schema (http://proatom.luksoft.cz/jaderneelektrarny/index.php?akce=reaktor&idtypbloku=6) (in Czech)
- Decommissioning of the reactor (http://www.iaea.org/inis/aws/htgr/abstracts/abst_29059905.html) (detailed report (http://www.iaea.org/inis/aws/htgr/fulltext/29059905.pdf), PDF)

Article Sources and Contributors

Bohunice_Nuclear_Power_Plant *Source*: http://en.wikipedia.org/w/index.php?title=Bohunice_Nuclear_Power_Plant *Contributors*: Atomique, Avala, Beagel, Bogdangiusca, Chris55, Corella, D6, Download, Dubhe.sk, Eumolpo, Grutness, Guyb123321, Hektor, Inwind, Iohannes Animosus, JanSuchy, Jimjamjak, John of Reading, Juro, Lysy, Mmarre, MosinNagant, Nazgul02, Pavel Vozenilek, Plastikspork, Rehman, Rjwilmsi, Sarvagyana guru, Tec15, Theanphibian, Waacstats, Whitepaw, Wizardist, Woohookitty, 29 anonymous edits

Nuclear_reactor *Source*: http://en.wikipedia.org/w/index.php?title=Nuclear_reactor *Contributors*: .:Ajvol:., 100DashSix, 130.94.122.xxx, 144.132.70.xxx, 1exec1, 2,4-dinitrophenylhydrazone, 216.10.184.xxx, 24.93.53.xxx, 2mcm, 777sms, A Softer Answer, A930913, ACSE, ANTIcarrot, Aarchiba, Abcdefgy2, Abledoc, Adhalanay, Admiral Roo, Adrianbg, Agisten, Ahoerstemeier, Albany NY, AlexHajnal, Aliensvortex, Allens, Altered13, Altes2009, AmiDaniel, AndrewHangChen, Andrewa, Andreww, Andy Dingley, Andycjp, AngelHedgie, Anlace, Antandrus, Arkuat, Asauers, Astronautics, Atarr, AxelBoldt, Axl, BD2412, BLuP1, Bart133, Barticus88, Baum, Benjamin Gatti, Beprepn, Bidabadi, Big Bob the Finder, BigFatBuddha, BilCat, BjörnBergman, Bkell, Black Kite, Blainster, Bluemoose, BobKawanaka, Bobblewik, Bobo192, Bongwarrior, Boundarylayer, Bped1985, BrightStarSky, Brock, Bryan Derksen, Bveazie, C-4, CDN99, Cactus.man, Calltech, Calpass, Cam, CanadianLinuxUser, Cancun771, Capricorn42, Captain panda, Carnildo, Celebration1981, Celestika6, Cffrost, Chealer, Chester Markel, Chitomcgee, Choplair, Chris G, ChrisGualtieri, Christian75, Christofer C. Bell, Ckatz, Clivefan, Closedmouth, Cmapm, Cmprince, Cobra12au, CobraWiki, Coenen, Cometstyles, Computerjoe, Conversion script, Cosmoskramer, Courcelles, Cquan, Crazysane, Cryptic, Cyrius, DARTH SIDIOUS 2, DJ Clayworth, DO'Neil, DV8 2XL, Dabbler, Dachshund, Damicatz, Dammit, Dan100, Daniel Quinlan, Daniel.Cardenas, Danielnez1, Davemck, David spector, David.Monniaux, DavidMIA, Defender of torch, Deglr6328, Demaag, Denni, DerHexer, Discospinster, Dreadstar, DrewT2, Dtcmcp99, Dtrcrtrc, Dusk7dragon, Dust of the world, Dwgriggs, Dylanisnoob, Długosz, EVERSON4LIFE, Eassin, Ed g2s, Edcolins, Eiland, Eilishsholai, El, Ellywa, Ember of Light, Emerson7, Emilio Juanatey, Emperorbma, Enr-v, Enzo Aquarius, Epbr123, Estonofunciona, Evercat, Everyking, Excirial, Exeunt, Fabiform, Fastfission, Favonian, FellGleaming, Ferkelparade, Finewinescotland, Fingers-of-Pyrex, Finlay McWalter, FocalPoint, Foobaz, Forteblast, Fred Bauder, Fredrik, FreplySpang, Frosted14, Funnybunny, Furrfu, Furrykef, Fuzheado, GD, GHe, Gaius Cornelius, Geljamin, Gene s, Gilliam, Gimeldaleth, Glenn, Gogo Dodo, Gopher65, Gordonfu, Goudzovski, GraemeL, Gralo, Gscshoyru, Gthb, Hadal, Haikupoet, HappyDog, Headbomb, Henry Flower, Heron, Hinklehoffer, Hmains, Hongooi, Htomfields, Hu12, Husnock, Hydro, Hydrogen Iodide, Hégésippe Cormier, Idiazabal26, Ignignot, Ikar.us, Ironboy11, Isnow, ItsZippy, IvanLanin, Ivolocy, Izehar, J.delanoy, JForget, JFreeman, JWB, JaGa, Jacob.jose, Jafet, Jaxl, Jean-François Clet, Ji.rodriguezmarin, Jim Cook, Jinglebb, Jmbeck, Joachim Wuttke, Joel7687, JohnOwens, Johnfos, Jonnabuz, Jorfer, Jorunn, Jozwolf, Jrthorpe, Jumping cheese, KFSorensen, Kanags, Kevin, Kflorek, Khazar2, Kheider, Kingturtle, Kjkolb, Kniwor, KoshVorlon, Kralizec!, Ksanyi, Kvng, LC, Lacrimosus, Lailaiboy, Larryzap, Lcolson, LeaW, Light show, Lightmouse, LikeaBLAWK, LilDice, Limulus, Linas, Linkman21, Linuxlad, LittleOldMe old, Llort, Logan, Loren36, LouriePieterse, Lumos3, Lupin, Lwnf360, LyeSolution, Mac, Malcolm Farmer, Man with two legs, Mark.murphy, Markmit, Matanya (renamed), Materialscientist, Matt Gies, Matusz, Maurice Carbonaro, Mav, Mel Etitis, Mentifisto, Mh26, Michael Hardy, MichaelHaeckel, Michfan2123, Mike Van Emmerik, Miken32, Millsdavid, Mindmatrix, Minimac's Clone, Mishlai, Mkweise, Mlk, Mormegil, Mr Minchin, MrOllie, Mschlindwein, Mtnerd, Mugget11, Musashi1600, Mvz34, N328KF, NPguy, Naddy, NaiPiak, Natalie Erin, Natpowning, Nealmcb, NerdyNSK, Neweb, Nickolay Stelmashenko, Nicransby, Nis81, Nixdorf, Nlilovic, Noctibus, NoiseEHC, Nomad, Nsaa, NuclearEnergy, Nyme, Ohnoitsjamie, Old Moonraker, Oldarney, Oldnoah, Oliverdl, OlivierM, Omegatron, Omicronpersei8, Opelio, Ophanna, Oralloy, Ortonmc, Otolemur crassicaudatus, Ottawakismet, Owneh, Oxymoron83, PMLawrence, Panoptik, Patrick, Paul August, Paul Raftery, Paxsimius, PeaceNT, Petri Krohn, Phidauex, Philip Trueman, Philipum, PhySusie, Pinethicket, Pip2andahalf, Piperh, PizzaofDoom, Playstationman, Pm504, Pmtanx, Polmandc, Polyamorph, Ponyo, Porqin, Prodego, Pseudomonas, Pstudier, PurpleChez, Quebec99, Quest for Truth, Quintote, Qxz, RG2, RJaguar3, RTC, RW Marloe, Rama, Ranveig, Ray Van De Walker, Razgriz6, Rbk, Rchandra, Rdsmith4, Read-write-services, Redlineeditor, Rei, Rememberway, Res2216firestar, Retired username, Rettetast, RexNL, Rich Farmbrough, Roadrunner, Robaker, Rod57, Ron Ritzman, Ronz, Rracecarr, Rrburke, Rsrikanth05, Rwendland, RxS, Ryulong, SDC, Samuel Blanning, Samussbrother, Samw, Sankalpdravid, Satellizer, Sbharris, Schutz, Scythia, SeanMack, Seanofthedead12, Seaphoto, Secfan, Seeleschneider, Seth Ilys, Shaddack, Shanes, ShotmanMaslo, Shrigley, Sietse Snel, Silent52, Simesa, Sitethief, Sk8ski, Skier Dude, SkyWalker, SnappingTurtle, Sodium, Soetermans, Speltcolor, Splatbargel, SriMesh, Srikeit, Sscott328, StarryGrandma, StaticGull, Stefancan, Steinsky, Stephenb, Stepho-wrs, Stereotek, Stevertigo, Stianl, Stripsense, Substar, Sudoaptitude, Suresh 5, Surv1v4l1st, Survivor, Susan Mason, Svenski666, Svick, Swarm, Swooshiain, Synchronism, Syrthiss, TMLutas, Tanthalas39, Tbhotch, Tdadamemd, Tdmidget, Team saint lucia, Teapeat, Tedd, Tedernst, Tempshill, That Guy, From That Show!, The Epopt, The High Fin Sperm Whale, The Random Editor, The Thing That Should Not Be, Theanphibian, Theda, Thegreenj, ThevikasIN, Thomas888b, Thunder8, Thunderbuck ram, Tiggerjay, TimVickers, Tobby72, TodKarlson, Tom harrison, Tomtheman5, Toytoy, Trelvis, Trevor MacInnis, Tricky Victoria, Trigor, Trojancowboy, Truthunmasked, Tuntable, Tweenk, Ultramarine, UncleDouggie, Uruiamme, Useight, Vanished user kjdioejh329io3rksdkj, Vasi31, Vegaswikian, Velveteman1, Vgy7ujm, Vibhijain, Vipinhari, Vrock25, Vsmith, Waggers, Waiper1, Wannabemodel, Warfieldian, Wavelength, Weregeek, Widr, WikiTyson, Wikianon, Wikieditoroftoday, Wikipelli, Wildnox, Wknight94, Wmahan, WpZurp, Wreck, Wtshymanski, Wwoods, XJaM, XLerate, Yash Vaidya, Yoonks995, Your Lord and Master, ZBrannigan, Zdv, Zereshk, Zigger, Zowie, Zslevi, ТимофейЛееСуда, 1014 anonymous edits

Jaslovské_Bohunice *Source*: http://en.wikipedia.org/w/index.php?title=Jaslovsk%C3%A9_Bohunice *Contributors*: ALadinN, BranoDD, D6, Dr. Blofeld, Frazzone, Juro, Kelovy, Makki98, MarkBA, Physchim62, Tankred, 1 anonymous edits

Pressurized_water_reactor *Source*: http://en.wikipedia.org/w/index.php?title=Pressurized_water_reactor *Contributors*: 999mal, A2Kafir, Aarchiba, Acq3, Ahpook, Alansohn, Alex.tan, Alfredo22, Alison, Alistair1978, Andrewa, Anthony Appleyard, Arjayay, Aronomy, Asauers, BarretB, Bejnar, Benedias, Bense, BigBen212, BilCat, Biscuittin, Blue387, Bobblewik, Bobo192, Bogey97, Br77rino, Brichcja, Brownh2o, Bryan Derksen, Buddha72, Caspertheghost, Chat1410, Chealer, ChrisJ, Christian.cantrell, Christopherfletcher870, Ciphers, Colinsweet, D6, DV8 2XL, Dan100, Darth Panda, David.Monniaux, Daviddavid, Dennyboy34, Dondon0, Dridgway, Drumguy8800, Dwp49423, EdH, Ederiel, Edward, Eiland, Electron9, Epbr123, Europrobe, Exsquid, Fastfission, Fsbr1908, G Clark, Gene Nygaard, Gorbyc, Gurch, Hektor, Hephaestos, Heron, Hmrox, Hqb, Hullernuc, Ianboggs, J.Ring, JLaTondre, JLasater, JWB, Jaufrec, JeremyR, Johnfos, Jonth, Kanadier, Keenan Pepper, Kestasjk, Kevin, Kevincar03, Knife Knut, Krich, Lamro, Lcolson, Lekrecteurmasque, Lewisdg2000, Li-sung, Lightmouse, Limulus, Lowellian, MBK004, Mackerm, Mashford, Michael Angelkovich, Michael Hardy, Mintleaf, Mion, Mkweise, MonteChristof, Morn, Mulad, Mwc06, NamSandStorm, Nbach, Northumbrian, Olivier Debre, Ortcutt, Ospalh, Ottawakismet, Pennywisdom2099, Petri Krohn, Phatom87, Piano non troppo, Piperh, Protonk, Puchiko, Quazgaa, R153nm, Rapatan, RedWolf, Revr J, Rjwilmsi, Rl, Roadrunner, Rocketboy50, Rod57, Rtcoles, Rwendland, SCARECROW, Saga City, Sandycx, ScottW, Sefarkas, ShakingSpirit, Shd, Simesa, Sleske, Snow Blizzard, Snrrub, Sohlemac, Sonnycnn, Stainless316, Stephan Leeds, Strangnet, Suisui, Suruena, TaintedMustard, Tantantantantan, Tarran, Technogreek43, The Rambling Man, Theanphibian, Theleek, ThurnerRupert, TitaniumCarbide, Tom verbeure, Tompw, Tvbanfield, Unconventional, Warut, Wdfarmer, Wendell, Widr, Wikipelli, Wizard191, Wknight94, Woohookitty, Wwoods, Yintan, Yuckfoo, Zakatan, Zoicon5, شروز, 338 anonymous edits

VVER *Source*: http://en.wikipedia.org/w/index.php?title=VVER *Contributors*: Adamrush, Andrewa, Arjayay, Arnoha, Asauers, Balcer, Beagel, Bender235, Berkut, BlackSlivers, Brhaspati, Brozen, C1010, Cassini83, ChNPP, Chaojoker, CommonsDelinker, Corella, Dust of the world, Dwalin, Enr-v, EugeneZelenko, GRindala, Gaius Cornelius, Hamishnz, Hibernian, Hullernuc, Huunta, Indochinetn, JanSöderback, Jkausel, John of Reading, Katana0182, Knowing guy, Lcolson, LilHelpa, Limulus, Lklundin, Los688, MPorciusCato, Marc Kupper, MarkBA, Midgley, Mikus, Mmarre, Nk, Nora lives, Ospalh, Petri Krohn, PhoenixHallow, Q43, Rjwilmsi, Rsmn, Rwendland, Simesa, Smithfarm, Tarduyabgu, Tec15, TheRingess, Thompn4, Timothy Cooper, Tom120, Tulkolahten, VargaA, Vmenkov, Xmort, Ysangkok, Васильев Виктор Николаевич, 112 anonymous edits

Electricity *Source*: http://en.wikipedia.org/w/index.php?title=Electricity *Contributors*: *andreweclark, 28421u2232nfenfcenc, 97198, ABF, Aaron Mendelson, Academic Challenger, Achmad Fahri, Acroterion, Addshore, Ahoerstemeier, Aidanpugh, Aim Here, Ak47225, Akasuna, Akorak, Akshay11, Akuyume, Alan16, Alansohn, Alex.tan, Alexandria, Alfred Centauri, Algumacoisaqq, Ali, Alphachimp, Alvaro.jaramillo, Alvinrune, Ameliorate!, AmericasPower, Amire80, Amitprabhakar, Ancheta Wis, Andres, Andrewpmk, Andrewthejesus, Andy12983, Angleofdeath, Anonymous101, Antandrus, Anthony Appleyard, Anticipation of a New Lover's Arrival, The, Antonio Lopez, Anwar saadat, Arch dude, Archer3, ArielGold, Aristolaos, ArmyHero, Askewchan, Asterion, Atomaton, AussieRulez, Average Earthman, Avono, AxelBoldt, AySz88, AzaToth, Azitnay, BBird, Barometer, Bart133, Barticus88, Bassbonerocks, Bawolff, Beagel, Beao, Beast of traal, Bekus, Beland, Bemoeial, BenFrantzDale, Bennyboys, Bensaccount, Bensin, Bergsten, Bert Hickman, Betacommand, Betterusername, BeyondHisYears, Beyondthislife, Bidabadi, BierHerr, Big G Ursa, BillC, Binary TSO, Biria64, Blair Bonnett, Blue520, Bluerfn, Boaster, Bobblewik, BobbyLee, Bobo192, Bogdangiusca, Bokkibear, Bong69, Bongwarrior, Bookermorgan, Brain40, Brandmeister (old), Brando130, Brian0918, Brockert, BrownBean, Bruce Foods, Bucephalus, Butros, C J Cowie, CA387, CAPS LOCK, CTF83!, CZero, CambridgeBayWeather, Cameron Dewe, Can't sleep, clown will eat me, Canderson7, CapitalR, Capricorn42, CaptainVindaloo, Carbon-16, Carlsotr, CarolGray, Casper2k3, Castncoot, Catgut, Cbornmann, Celebration1981, CesarFelipe, Chadmilerna, Chairboy, Children.of.the.Kron, Choiboi22, ChrisGualtieri, Chrislk02, ChristopherEdwards, Christophercm, Citicat, Cjohn67, Ckatz, Clappingsimon, CliffordWest, Closedmouth, Clw, Cockman45, Cockman46, Codice1000.en, Coemgenus, Cometstyles, Conversion script, Coolblaze, Covington, Crezelnuts93, Ctbolt, Cuddlyopedia, Cyclonenim, DV8 2XL, DVD R W, Dacium, Dagnabit, Dalyman, Damicatz, Daniel.Cardenas, Danielkwalsh, Dar-Ape, DarkFalls, Darkwind, Datta.naikwade, Davewild, DavidLevinson, DavidSaff, Dbachmann, Dbfirs, Dcpc0807, DeLarge, Deadcode, Debresser, Dedbaby, Dekimasu, Deli nk, Delicious carbuncle, Delirium, Delldot, Deltabeignet, Demiurge, Dendodge, DerHexer, Derek Ross, Dictyosiphonaceae, Dina, Dingy, Disavian, Discospinster, Dkeeper, Doc glasgow, DocWatson42, Dogears, Dominick, DonVander, Doomavenger1, DoubleBlue, Dreadstar, Drew2131, Drostie, Ds13, Dsdickie, Dtrak, Dual Freq, DumLoco, Durin, Dwilso, Dysepsion, EHRice, EarthPerson, Edgar181, Edgarde, Edivorce, Ee00224, Eengined, Egil, Eisd, Elaragirl, Elassint, Electriccatfish2, Elfguy, Eliz81, Ellbeecee, Elmicker, Eltanin, Embrittled, Epbr123, Equendil, Er Komandante, Erebus Morgaine, Eric-Wester, EscapingLife, Esn, Esoltas, Ethan, Eubulide, EugeneZelenko, Everyking, Evrik, Fasten, Fauzan, Femto, Fieldday-sunday, Filtay, Fir0002, Fizicist, Flewis, Florian Blaschke, Fluri, Foraneagle2, Fredbauder, FreplySpang, Freshacconci, Fruitytingles, Fuhghettaboutit, FuturedOrange, G.michalis, Gaius Cornelius, Gecko, Geneb1955, Geomanjo, Gerry Ashton, Ghou, Giftlite, Gilliam, Glane23, Glaze012, Glenn, GliderMaven, Gobonobo, GodIsGreater, Goethean, Gogo Dodo, Gottadmit, GraemeL, Graham87, Gralo, GraybeardThePirot, Grendelkhan, Gtxfrance, Gurch, Hadal, Hadrians, Haham hanuka, Hallenrm, Happinessiseasy, Harland1, Harryzilber, Hbiz777, Hdt83, Headbomb, Heimstern, Henrik, HenryLi, Henrygb, Heron, Hhhippo, Hiberniantears, Hifrommike65, Hiphopdancequeen, Hipocrite, Hippi ippi, Hispalois, HonztheBusDriver, Hotdogmckgee, HugoLoris, Hulek, Hveziris, HybridBoy, IRP, IW.HG, Iain99, Iames, Icairns, IceUnshattered, Igjav, Imasleepviking, Imjustmatthew, Imnotminkus, ImperfectlyInformed, Improbcat, Inluminetuovidebimuslumen, InvertRect, Iridescent, Irishguy, Iritebs, Iritebs1, Itai, Ivan Bajlo, Ixfd64, J. W. Love, J.delanoy, J04n, J8079s, JDLukas, JHeinonen, JHunterJ, JSquish, Jackehammond, JackyR, Jagged 85, Jaknouse, Jarekadam, Jason Patton, JayTau, Jb.schneider-electric, Jc3s5h, Jdolan2726, Jeblo, Jedibob5, Jeff hardy3456, Jellevc, Jerry, Jerzy, Jhinman, Jimmyx90, JinJian, Jmlk17, Jni, JodyB, JoeSmack, Joeylawn, John Reaves, John Stumbles, Jolibrarian, JonDon69,

Jorge Ianis, Jose77, Joseph Banks, Josh Grosse, Joshua Scott, Jpgordon, Jrockley, Jsd, Jshadias, JuJube, Juliancolton, Justin Eiler, Justinritter, Jwissick, KCinDC, Kablammo, Karafias, Karl2620, Karlawilloughby, Katalaveno, Keilana, Kelly190, Kerotan, KerryO77, KevinTR, Khalid Mahmood, Khukri, King Bowser 64, King Toadsworth, King of Hearts, Kitty53, Kizor, Klemen Kocjancic, KnowledgeOfSelf, Kolakowski, Koliri, Koolyman, Korg, Koyaanis Qatsi, KrakatoaKatie, Kramden4700, Kramer, Krash, Kubigula, Kungfuadam, Kuru, L33th4x0rguy, LOL, Lagelspeil, Latka, LeaveSleaves, Leffmann, Lemonflash, Leon7, Levineps, Lichen from Hell, Light current, LilHelpa, Lily15, Lineface, Linuxmatt, LittleOldMe, Lmatt, Looxix, Lostkey2, Lradrama, Luigifan, Lumberjack steve, Luminate, Lumos3, Luna Santin, M C Y 1008, M bastow, M6060, MBisanz, MC MasterChef, MER-C, MONGO, Mac, Machoman739, Madsmokey, Magnagr, Maias, Malerin, Man vyi, Mani1, Marcus Qwertyus, Marek69, Marj Tiefert, Mark Dingemanse, MarkS, MarkSutton, Markgriz, MarsRover, Martin451, Masterchamp2, Matsuiny2004, Matt Deres, Mattarata, Matthead, Matthuxtable, Matticus78, Mattyh190, Maximus Rex, Mbstone, McSly, Mdebets, Me the Third, Megan.rw1, Mejor Los Indios, Melca, Mexicanpunjab, Michael Hardy, Michael Snow, Miguelaaron, Mike Rosoft, Mike V, Mike5904, MikeLeeds, Mild Bill Hiccup, Mintrick, Mion, Miquonranger03, Mirwin, Mishkin11, Mishuletz, Mitch Ames, Mjpieters, Mman33, Monedula, Monkeyman, Moofy, Morrowulf, Mpa, Mr Mulliner, MsVanAuken, Mschel, Mszegedy, Muffinstastenice, Mulad, Munci, Muysal, Myanw, Myasuda, Mygerardromance, N1RK4UDSK714, NAHID, NHRHS2010, Nabla, Nader85021, Nakon, Namazu-tron, Ncmvocalist, Nebather, Needtoknow31, NeilN, NeoAdonis, Neoguy9090, Netscott, Neverquick, NewEnglandYankee, Nibuod, Nick, Nick C, Nick0416, Nickdog33, Nixdorf, Nmnogueira, Noctibus, Nojika, Nposs, Nsaa, NuclearEnergy, Nukeless, Nurg, Nwwaew, Oamaro, Oda Mari, Ohnoitsjamie, Oleg Alexandrov, Oli Filth, Omegatron, Onebravemonkey, Onorem, OrbitOne, Oreo Priest, Oroso, Oscarthecat, Oshwah, Ot, Oxymoron83, P3rs0n, PCHS-NJROTC, Pahles, Palmerston, PandoraX, Papa November, Patrick, Patrolboat, Paul-L, Pauldavidgill, Pax:Vobiscum, PedEye1, Pedro, Persian Poet Gal, PeterCanthropus, Peterlin, Petri Krohn, Pgk, Phileas, Philip Trueman, PhilipMW, Picapica, Piesaretasty345, Pinkadelica, Piperh, Planetscared, Platyk, Poccil, PoccilScript, Polluxian, Pomte, Porterjoh, Prabhat278205, Prashanthns, Praveenp, PrestonH, Prettyponies12343, Primasz, Primetime, Private Butcher, Promethean, PrometheusDesmotes, Proofreader77, Puckett34, Pucketteli, Pyr, Pyrospirit, Quadrius, Quiddity, Quintote, Qxz, Qyd, R'n'B, RHJesusFreak40, RJaguar3, RTC, RaAnubisOsiris, Radon210, Rahulvohra29, RainbowOfLight, Randolph02, Randomblue, Ranveig, Rasmus Faber, Raul654, Rauljdelgado, Raven4x4x, Rawling, Rc3784, Rd232, Rdsmith4, Reconday41, Reddi, Remshad, Repersplitter, ResidentAnthropologist, Retiono Virginian, Rettetast, RexNL, Rheostatik, Riana, Rich Farmbrough, Rjstott, Rjwilmsi, Rob Hooft, Robert K S, Robert Merkel, Robin Hood, RockMFR, Rogator, Room214, RoyBoy, Rrburke, Rundquist, Ruud Koot, Ruyn, RxS, Ryan, Ryan032, Ryanswimstheworld, Ryguasu, Ryguillian, SCEhardt, ST47, SWAdair, Sadi Carnot, Saenzc, Saga City, Sam.P.Hollins, Samjohnson, Sammetsfan, Sandstein, Sarah, Saulicious, Sayeth, Sb617, Sbharris, Sbyrnes321, Sceptre, Scetoaux, Schwnj, ScienceMind, Scientizzle, Scottfisher, Seazoleta, Sf5xeplus, Shadow1, ShakingSpirit, Shanedidona, Shangrilaista, Sharkie3000, Shattered, SheepNotGoats, Shell Kinney, Shenme, Shizane, Shoedizzog, Sietse Snel, Sifaka, Silverstein210, Singularity, Sir Vicious, SirGrant, Sitethief, Sjakkalle, Sjcandyman, Skeledzija, Skizzik, Skunkboy74, SkyWalker, Slash, SlickWillyLovesSex, SlimVirgin, Smalljim, Smark33021, Smellmyfeet, Smitz, SnappingTurtle, Snozzer, Someguy1221, Soundoftheunderground, SparrowsWing, Squids and Chips, Srleffler, Ssd, Ssilvers, Stbalbach, Stemonitis, Stephenb, Steve Pucci, Steve0913, Steven Zhang, Steventrouble, Stevenwrose, Stevertigo, Storm Rider, Stubes99, Sub zero133, Subash.chandran007, Superbeecat, Surfin simo, SusanLesch, Swiftblade6, Symane, Synflame, Syrthiss, Szczureq, Szlam, T, Tariqabjotu, Tarret, Tassedethe, Tawker, Tbone762, Teapeat, Technopat, Ted Longstaffe, Teemu Ruskeepää, TehNomad, Template namespace initialisation script, TexasAndroid, The Founders Intent, The Rambling Man, The Thing That Should Not Be, The X, The undertow, The wub, Theda, Theunicyclegirl, Thewob, Thingg, Thisisbossi, Tholme, Thrissel, Tiles, Tim Starling, Timc, Timclare, Titoxd, Tjdw, Tkynerd, Tlabshier, Tomahiv, Tomasz Prochownik, Tombomp, Tony Fox, Tpth, Treisijs, Ttwaring, Tubbyalonso, Tulls55, Tylko, Uarshad82, Ubcule, UberScienceNerd, UkPaolo, Unschool, Useight, User2301, Usnerd, Utcursch, Vald, Vanished user 9i39j3, VashiDonsk, Vedexent, Velho, Vsmith, WAvegetarian, WODUP, WarthogDemon, WatermelonPotion, Waveguy, Wavelength, Waycool27, Wayward, Welzen, Wheeeeee, Whispering, Whitepaw, Why Not A Duck, Widefox, Wiki alf, William Avery, Willy56.5, Wimt, Wjbeaty, Wknight94, Wongm, Woohookitty, Wtshymanski, Wutime, Wwoods, X!, X-Tron 13, X2RADialbomber, Xaosflux, Xerxes314, Xev lexx, Xnuala, Xwu, Y, Yamamoto Ichiro, Yekrats, Yickbob, Ying123, Yjwong, Yobmod, Youngster14, Youssefsan, Zaingay, Zalgo, Zanthrax, Zapvet, Zarathura, Zelikazi, Zigger, Zoicon5, Zoragotcha, Zrogerz69, Zslevi, Zudduz, Zundark, Zureks, Zuwiki, , 1848 anonymous edits

Cogeneration *Source*: http://en.wikipedia.org/w/index.php?title=Cogeneration *Contributors*: Adamrush, Alan Liefting, Amberger, Americasroof, Andres, Animagi1981, Antonin, Armando, ArnoldReinhold, Article editor, BLueFiSH.as, Bava Alcide57, Beagel, Beetstra, Beland, Billymac00, Biomanuk, Blehfu, Bogdan 007, Bsadowski1, CHPfactchecker, CRGreathouse, CanadianPenguin, Cavrdg, Cgingold, Chp109, Chpfocus, Cirt, Ckatz, Claush66, Crosbiesmith, CumbiaDude, Delusion23, Dgdeckard, Dispenser, Djdiablo79, Drestros power, Druzhnik, Dtobias, Eeekster, El C, Elekhh, Enercon Engineering Inc., Energee5, Enerjiturk, Engineman, Erodium, Espoo, Essicajay, Euchiasmus, Fahidka, Farpointer, Fletcher, Folifo, Foofighter20x, GGByte, Gdnumber1, Giftlite, Gralo, Gueneverey, Gunnar.Kaestle, Hamm1344, Helloitsonlyme, Hmains, Howard Hall Farm, Hu12, Intern8, Ioeth, J.delanoy, JDspeeder1, JEdlund, Jamesmorrison, Jan1nad, Jensbn, Jespernoes, Jetam2, Jjjbell, Jmax-, Joeknight1, Jojalozzo, Jzoch2, Kizor, Koavf, Krzysiu Jarzyna, Lawrencekhoo, Leifhille, LilHelpa, Linas, MGTom, Mako098765, Mandarax, Markaci, Mattcguy, Mattia Luigi Nappi, Midgley, Mion, Mnmngb, Moe Epsilon, Mogism, Monkofthetrueschool, Mortense, Mostafa9971, MuddyDave, Munay09, NeilN, Niceguyedc, Northamerica1000, Oakmedia, Ohnoitsjamie, OilAndGasExpert, Old Moonraker, Oliver.gouldthorpe, Onthetopofthehill, Otduff, Ouro, Pahazzard, Pakaraki, Pearle, Pege.founder, Pekiro, Peterlewis, Philip Trueman, Phlow1979, Phmoreno, Pinethicket, Propagator, Qonline, Ray Van De Walker, Rcandamoonpie, Rd232, Reaper Eternal, Redrose64, Rex the first, Rich Farmbrough, Rich257, Richtman, Rjwilmsi, Rmhermen, Roadrunner, RobertG, Rocketrod1960, Rockfang, RoyBoy, Salix alba, Samw, Sandpiper, SchreiberBike, ScottSteiner, Season21, Sfreyer, ShadowInfernoBlitz, Shore3, Skobayas, Slambo, Smack, Smithfarm, Speakforthose, Stickee, Sylvain.quoilin, TAXman, TSP, Teratornis, Thameshead, Timeshifter, TolarisTango, User A1, Vegas949, Vegaswikian, Vmenkov, Vortexrealm, Vsmith, Wendydi, Wolfkeeper, Wtshymanski, Wykis, Yeliseyev, Zenibus, Zyads, علی ویی, 225 anonymous edits

KS_150 *Source*: http://en.wikipedia.org/w/index.php?title=KS_150 *Contributors*: Abdull, Airplaneman, Andycjp, Corella, Eumolpo, Felipe.sanches, Hugo999, Kinu, Lan Di, Maralia, Neelix, Pavel Vozenilek, Rod57, Tec15, Wavelength, Wtshymanski, 15 anonymous edits

Image Sources, Licenses and Contributors

File:Crocus-p1020491.jpg *Source*: http://en.wikipedia.org/w/index.php?title=File:Crocus-p1020491.jpg *License*: unknown *Contributors*: User:Rama

File:Nuclear fission.svg *Source*: http://en.wikipedia.org/w/index.php?title=File:Nuclear_fission.svg *License*: unknown *Contributors*: User:Fastfission

File:Otto Hahn und Lise Meitner.jpg *Source*: http://en.wikipedia.org/w/index.php?title=File:Otto_Hahn_und_Lise_Meitner.jpg *License*: unknown *Contributors*: unbekannt

File:ChicagoPileTeam.png *Source*: http://en.wikipedia.org/w/index.php?title=File:ChicagoPileTeam.png *License*: unknown *Contributors*: Ary29, Bomazi, Clever Hans, Davepape, DroEsperanto, Fastfission, Hawkeye7, Infrogmation, Mu, Mutter Erde, Pieter Kuiper, Wutsje, 2 anonymous edits

File:Pulstar1.jpg *Source*: http://en.wikipedia.org/w/index.php?title=File:Pulstar1.jpg *License*: unknown *Contributors*: Porao, 1 anonymous edits

File:Pulstar2.jpg *Source*: http://en.wikipedia.org/w/index.php?title=File:Pulstar2.jpg *License*: unknown *Contributors*: Original uploader was Zereshk at en.wikipedia

File:RIAN archive 450312 Treatment of interior part of reactor frame.jpg *Source*: http://en.wikipedia.org/w/index.php?title=File:RIAN_archive_450312_Treatment_of_interior_part_of_reactor_frame.jpg *License*: unknown *Contributors*: A. Khrupov / A. Хрупов

File:Thermal reactor diagram.png *Source*: http://en.wikipedia.org/w/index.php?title=File:Thermal_reactor_diagram.png *License*: unknown *Contributors*: Aarchiba, Bomazi, Dcoetzee, Duesentrieb, EugeneZelenko, Nuno Tavares, Slashme, Theanphibian, Was a bee

File:Diablo canyon nuclear power plant.jpg *Source*: http://en.wikipedia.org/w/index.php?title=File:Diablo_canyon_nuclear_power_plant.jpg *License*: unknown *Contributors*: Billinghurst, FlickrLickr, FlickreviewR, KaHe, Look2See1, Mrrxx, Nolanus, 2 anonymous edits

File:Laguna Verde Nuclear Power Plant.jpg *Source*: http://en.wikipedia.org/w/index.php?title=File:Laguna_Verde_Nuclear_Power_Plant.jpg *License*: unknown *Contributors*: Original uploader was Exarkunmx at en.wikipedia Later versions were uploaded by Theanphibian at en.wikipedia.

File:CANDU at Qinshan.jpg *Source*: http://en.wikipedia.org/w/index.php?title=File:CANDU_at_Qinshan.jpg *License*: unknown *Contributors*: AECL

File:Elektrownia Ignalina.jpg *Source*: http://en.wikipedia.org/w/index.php?title=File:Elektrownia_Ignalina.jpg *License*: unknown *Contributors*: User:Julo

File:Sizewell A.jpg *Source*: http://en.wikipedia.org/w/index.php?title=File:Sizewell_A.jpg *License*: unknown *Contributors*: Bomazi, Fintan264, FrancisTyers, Geni, Oneblackline, Theanphibian

File:Torness Nuclear Power Station, Scotland.JPG *Source*: http://en.wikipedia.org/w/index.php?title=File:Torness_Nuclear_Power_Station,_Scotland.JPG *License*: unknown *Contributors*: (also unused "Improved contrast" version by , reverted by)

File:Superphénix.jpg *Source*: http://en.wikipedia.org/w/index.php?title=File:Superphénix.jpg *License*: unknown *Contributors*: Yann Forget

File:Fukushima I by Digital Globe.jpg *Source*: http://en.wikipedia.org/w/index.php?title=File:Fukushima_I_by_Digital_Globe.jpg *License*: unknown *Contributors*: Digital Globe

Image: Slovakia - background map.png *Source*: http://en.wikipedia.org/w/index.php?title=File:Slovakia_-_background_map.png *License*: unknown *Contributors*: Caroig

Image: Slovakia - outline map.svg *Source*: http://en.wikipedia.org/w/index.php?title=File:Slovakia_-_outline_map.svg *License*: unknown *Contributors*: Caroig

File:Red pog.svg *Source*: http://en.wikipedia.org/w/index.php?title=File:Red_pog.svg *License*: unknown *Contributors*: Anomie

Image: Trnava Region - background map.png *Source*: http://en.wikipedia.org/w/index.php?title=File:Trnava_Region_-_background_map.png *License*: unknown *Contributors*: Caroig

Image: Trnava Region - outline map.svg *Source*: http://en.wikipedia.org/w/index.php?title=File:Trnava_Region_-_outline_map.svg *License*: unknown *Contributors*: Caroig

Image:Reactor Vessel head.jpg *Source*: http://en.wikipedia.org/w/index.php?title=File:Reactor_Vessel_head.jpg *License*: unknown *Contributors*: NRC

File:PressurizedWaterReactor.gif *Source*: http://en.wikipedia.org/w/index.php?title=File:PressurizedWaterReactor.gif *License*: unknown *Contributors*: U.S.NRC.

Image:Reactorvessel.gif *Source*: http://en.wikipedia.org/w/index.php?title=File:Reactorvessel.gif *License*: unknown *Contributors*: ChNPP, Flappiefh, Joelholdsworth, Karelj, Lcolson, Nolanus

Image:Nuclear fuel element.jpg *Source*: http://en.wikipedia.org/w/index.php?title=File:Nuclear_fuel_element.jpg *License*: unknown *Contributors*: PD-USGov

File:Wwer-1000-scheme.png *Source*: http://en.wikipedia.org/w/index.php?title=File:Wwer-1000-scheme.png *License*: unknown *Contributors*: User:Panther

File:Сравнение активных зон реакторов.svg *Source*: http://en.wikipedia.org/w/index.php?title=File:Сравнение_активных_зон_реакторов.svg *License*: unknown *Contributors*: Das steinerne Herz, Hullernuc

File:VVER-1000-Stereometric.svg *Source*: http://en.wikipedia.org/w/index.php?title=File:VVER-1000-Stereometric.svg *License*: unknown *Contributors*: User:RusCloud

File:Loviisan voimalaitos ilmasta.png *Source*: http://en.wikipedia.org/w/index.php?title=File:Loviisan_voimalaitos_ilmasta.png *License*: unknown *Contributors*: A333, Apalsola, Estormiz, MPorciusCato, Rwendland

File:Kozloduy Nuclear Power Plant - Control Room of Unit 5.jpg *Source*: http://en.wikipedia.org/w/index.php?title=File:Kozloduy_Nuclear_Power_Plant_-_Control_Room_of_Unit_5.jpg *License*: unknown *Contributors*: Yovko Lambrev

Image:BalakovoNPP3.jpg *Source*: http://en.wikipedia.org/w/index.php?title=File:BalakovoNPP3.jpg *License*: unknown *Contributors*: Александр Ситенький (Alexander Seetenky)

File:Flag of Russia.svg *Source*: http://en.wikipedia.org/w/index.php?title=File:Flag_of_Russia.svg *License*: unknown *Contributors*: Anomie, Zscout370

Image:Ровенская АЭС блок 4.JPG *Source*: http://en.wikipedia.org/w/index.php?title=File:Ровенская_АЭС_блок_4.JPG *License*: unknown *Contributors*: Victor Korniyenko

File:Flag of Ukraine.svg *Source*: http://en.wikipedia.org/w/index.php?title=File:Flag_of_Ukraine.svg *License*: unknown *Contributors*: User:Jon Harald Søby, User:Zscout370

Image:Khmelnytskyi Nuclear Power Plant Unit No.jpg *Source*: http://en.wikipedia.org/w/index.php?title=File:Khmelnytskyi_Nuclear_Power_Plant_Unit_No.jpg *License*: unknown *Contributors*: User:VargaA

Image:BalakovoNPP CR.jpg *Source*: http://en.wikipedia.org/w/index.php?title=File:BalakovoNPP_CR.jpg *License*: unknown *Contributors*: Александр Ситенький (Alexander Seetenky)

Image:Leningrad Nuclear Power Plant 20JUL2010-4 (cropped).jpg *Source*: http://en.wikipedia.org/w/index.php?title=File:Leningrad_Nuclear_Power_Plant_20JUL2010-4_(cropped).jpg *License*: unknown *Contributors*: User:Bayoue, User:Saibo

Image:Novovoronezh Nuclear Power Plant II-1.JPG *Source*: http://en.wikipedia.org/w/index.php?title=File:Novovoronezh_Nuclear_Power_Plant_II-1.JPG *License*: unknown *Contributors*: User:Hullernuc

Image:Novovoronezh Nuclear Power Plant II-CT.JPG *Source*: http://en.wikipedia.org/w/index.php?title=File:Novovoronezh_Nuclear_Power_Plant_II-CT.JPG *License*: unknown *Contributors*: User:Hullernuc

Image:Kudankulam NPP.jpg *Source*: http://en.wikipedia.org/w/index.php?title=File:Kudankulam_NPP.jpg *License*: unknown *Contributors*: Petr Pavlicek/IAEA

File:Flag of India.svg *Source*: http://en.wikipedia.org/w/index.php?title=File:Flag_of_India.svg *License*: unknown *Contributors*: Anomie, Mifter

Image:Kk stendal.jpg *Source*: http://en.wikipedia.org/w/index.php?title=File:Kk_stendal.jpg *License*: unknown *Contributors*: Original uploader was Nicransby at de.wikipedia (Original text : Nic Ransby)

File:Flag of Germany.svg *Source*: http://en.wikipedia.org/w/index.php?title=File:Flag_of_Germany.svg *License*: unknown *Contributors*: Anomie

Image:Ejz bloki.jpg *Source*: http://en.wikipedia.org/w/index.php?title=File:Ejz_bloki.jpg *License*: unknown *Contributors*: ChNPP, Masur, Piotrus, Powolniak, Theanphibian, Wpedzich, 2 anonymous edits

File:Flag of Poland.svg *Source*: http://en.wikipedia.org/w/index.php?title=File:Flag_of_Poland.svg *License*: unknown *Contributors*: Anomie, Mifter

Image:Juragua Nuclear Power Plant-2.jpg *Source*: http://en.wikipedia.org/w/index.php?title=File:Juragua_Nuclear_Power_Plant-2.jpg *License*: unknown *Contributors*: User:Felix Koenig

File:Flag of Cuba.svg *Source*: http://en.wikipedia.org/w/index.php?title=File:Flag_of_Cuba.svg *License*: unknown *Contributors*: see below

Image:Bundesarchiv Bild 183-1990-0221-029, Greifswald, Störfall im Kernkraftwerk.jpg *Source*: http://en.wikipedia.org/w/index.php?title=File:Bundesarchiv_Bild_183-1990-0221-029,_Greifswald,_Störfall_im_Kernkraftwerk.jpg *License*: unknown *Contributors*: Duch.seb, Video2005, Väsk

File:Lightning3.jpg *Source*: http://en.wikipedia.org/w/index.php?title=File:Lightning3.jpg *License*: unknown *Contributors*: U.S. Air Force photo by Edward Aspera Jr.

File:Thales.jpg *Source*: http://en.wikipedia.org/w/index.php?title=File:Thales.jpg *License*: unknown *Contributors*: Bibi Saint-Pol, Tomisti

File:Franklin-Benjamin-LOC.jpg *Source*: http://en.wikipedia.org/w/index.php?title=File:Franklin-Benjamin-LOC.jpg *License*: unknown *Contributors*: Androstachys, Davepape, Ecummenic, Editor at Large, Eubulides, Jengod, Jh12, Kilom691, Mschel, Nagy, Nonenmac, Raymond, Shizhao, 3 anonymous edits

File:M Faraday Th Phillips oil 1842.jpg *Source*: http://en.wikipedia.org/w/index.php?title=File:M_Faraday_Th_Phillips_oil_1842.jpg *License*: unknown *Contributors*: Thomas Phillips

File:Electroscope.svg *Source*: http://en.wikipedia.org/w/index.php?title=File:Electroscope.svg *License*: unknown *Contributors*: User:Beao, User:Stw

File:Lichtbogen 3000 Volt.jpg *Source*: http://en.wikipedia.org/w/index.php?title=File:Lichtbogen_3000_Volt.jpg *License*: unknown *Contributors*: User:Achgro

Image:VFPt image charge plane horizontal.svg *Source*: http://en.wikipedia.org/w/index.php?title=File:VFPt_image_charge_plane_horizontal.svg *License*: unknown *Contributors*: User:Geek3

File:Panasonic-oxyride.jpg *Source*: http://en.wikipedia.org/w/index.php?title=File:Panasonic-oxyride.jpg *License*: unknown *Contributors*: User:jaques

File:Electromagnetism.svg *Source*: http://en.wikipedia.org/w/index.php?title=File:Electromagnetism.svg *License*: unknown *Contributors*: User:Stannered

File:Electric motor cycle 3.png *Source*: http://en.wikipedia.org/w/index.php?title=File:Electric_motor_cycle_3.png *License*: unknown *Contributors*: Dmitry G, Isi, Tano4595

File:Ohms law voltage source.svg *Source*: http://en.wikipedia.org/w/index.php?title=File:Ohms_law_voltage_source.svg *License*: unknown *Contributors*: Kilom691, Ma-Lik, Omegatron

File:Gorskii 04414u.jpg *Source*: http://en.wikipedia.org/w/index.php?title=File:Gorskii_04414u.jpg *License*: unknown *Contributors*: digital rendering for the Library of Congress by Walter Frankhauser / WalterStudio

File:Parque eólico La Muela.jpg *Source*: http://en.wikipedia.org/w/index.php?title=File:Parque_eólico_La_Muela.jpg *License*: unknown *Contributors*: AnRo0002, Apalsola, Teratornis, Willtron

Image:Gluehlampe 01 KMJ.png *Source*: http://en.wikipedia.org/w/index.php?title=File:Gluehlampe_01_KMJ.png *License*: unknown *Contributors*: User:Edokter

File:The 1 Train.jpg *Source*: http://en.wikipedia.org/w/index.php?title=File:The_1_Train.jpg *License*: unknown *Contributors*: User:Sammetsfan

Image:Electric-eel2.jpg *Source*: http://en.wikipedia.org/w/index.php?title=File:Electric-eel2.jpg *License*: unknown *Contributors*: User:Stevenj

File:Trigeneration Cycle.jpg *Source*: http://en.wikipedia.org/w/index.php?title=File:Trigeneration_Cycle.jpg *License*: unknown *Contributors*: Jonnathan McForlan

Image:Masnedø power station.jpg *Source*: http://en.wikipedia.org/w/index.php?title=File:Masnedø_power_station.jpg *License*: unknown *Contributors*: Nillerdk, Sandpiper

File:Metz_biomass_power_station.jpg *Source*: http://en.wikipedia.org/w/index.php?title=File:Metz_biomass_power_station.jpg *License*: unknown *Contributors*: User:Bava Alcide57

File:Power plant at sunset.jpg *Source*: http://en.wikipedia.org/w/index.php?title=File:Power_plant_at_sunset.jpg *License*: unknown *Contributors*: User:Mattia Luigi Nappi

File:Mirant Kendall Cogeneration Station.jpg *Source*: http://en.wikipedia.org/w/index.php?title=File:Mirant_Kendall_Cogeneration_Station.jpg *License*: unknown *Contributors*: User:Fletcher6

MIX
Papier aus verantwortungsvollen Quellen
Paper from responsible sources
FSC® C105338

Printed by Books on Demand GmbH, Norderstedt / Germany